CAMBRIDGE STUDIES IN
ADVANCED MATHEMATICS 90

GLOBAL METHODS FOR COMBINATORIAL ISOPERIMETRIC PROBLEMS

Already published; for full details see http://publishing.cambridge.org./stm/mathematics/csam/

11 J.L. Alperin *Local representation theory*
12 P. Koosis *The logarithmic integral I*
13 A. Pietsch *Eigenvalues and s-numbers*
14 S.J. Patterson *An introduction to the theory of the Riemann zeta-function*
15 H.J. Baues *Algebraic homotopy*
16 V.S. Varadarajan *Introduction to harmonic analysis on semisimple Lie groups*
17 W. Dicks & M. Dunwoody *Groups acting on graphs*
18 L.J. Corwin & F.P. Greenleaf *Representations of nilpotent Lie groups and their applications*
19 R. Fritsch & R. Piccinini *Cellular structures in topology*
20 H. Klingen *Introductory lectures on Siegel modular forms*
21 P. Koosis *The logarithmic integral II*
22 M.J. Collins *Representations and characters of finite groups*
24 H. Kunita *Stochastic flows and stochastic differential equations*
25 P. Wojtaszczyk *Banach spaces for analysts*
26 J.E. Gilbert & M.A.M. Murray *Clifford algebras and Dirac operators in harmonic analysis*
27 A. Frohlich & M.J. Taylor *Algebraic number theory*
28 K. Goebel & W.A. Kirk *Topics in metric fixed point theory*
29 J.E. Humphreys *Reflection groups and Coxeter gropus*
30 D.J. Benson *Representations and cohomology I*
31 D.J. Benson *Representations and cohomology II*
32 C. Allday & V. Puppe *Cohomological methods in transformation groups*
33 C. Soulé et al *Lectures on Arakelov geometry*
34 A. Ambrosetti & G. Prodi *A primer of nonlinear analysis*
35 J. Palis & F. Takens *Hyperbolicity, stability and chaos at homoclinic bifurcations*
37 Y. Meyer *Wavelets and operators I*
38 C. Weibel *An introduction to homological algebra*
39 W. Bruns & J. Herzog *Cohen-Macaulay rings*
40 V. Snaith *Explicit Brauer induction*
41 G. Laumon *Cohomology of Drinfield modular varieties I*
42 E.B. Davies *Spectral theory and differential operators*
43 J. Diestel, H. Jarchow & A. Tonge *Absolutely summing operators*
44 P. Mattila *Geometry of sets and measures in euclidean spaces*
45 R. Pinsky *Positive harmonic functions and diffusion*
46 G. Tenenbaum *Introduction to analytic and probabilistic number theory*
47 C. Peskine *An algebraic introduction to complex projective geometry I*
48 Y. Meyer & R. Coifman *Wavelets and operators II*
49 R. Stanley *Enumerative combinatorics I*
50 I. Porteous *Clifford algebras and the classical groups*
51 M. Audin *Spinning tops*
52 V. Jurdjevic *Geometric control theory*
53 H. Voelklein *Groups as Galois groups*
54 J. Le Potier *Lectures in vector bundles*
55 D. Bump *Automorphic forms*
56 G. Laumon *Cohomology of Drinfeld modular varieties II*
57 D.M. Clarke & B.A. Davey *Natural dualities for the working algebraist*
59 P. Taylor *Practical foundations of mathematics*
60 M. Brodmann & R. Sharp *Local cohomology*
61 J.D. Dixon, M.P.F. Du Sautoy, A. Mann & D. Segal *Analytic pro-p groups, 2nd edition*
62 R. Stanley *Enumerative combinatorics II*
64 J. Jost & X. Li-Jost *Calculus of variations*
68 Ken-iti Sato *Lévy processes and infinitely divisible distributions*
71 R. Blei *Analysis in integer and fractional dimensions*
72 F. Borceux & G. Janelidze *Galois theories*
73 B. Bollobás *Random graphs*
74 R.M. Dudley *Real analysis and probability*
75 T. Sheil-Small *Complex polynomials*
76 C. Voisin *Hodge theory and complex algebraic geometry I*
77 C. Voisin *Hodge theory and complex algebraic geometry II*
78 V. Paulsen *Completely bounded maps and operator algebras*
79 F. Gesztesy & H. Holden *Soliton equations and their algebro geometric solutions I*
81 S. Mukai *An introduction to invariants and moduli*
82 G. Toutlakis *Lectures in logic and set theory I*
83 G. Toutlakis *Lectures in logic and set theory II*
84 R.A. Bailey *Association schemes*
91 I. Moerdijk & J. Mrcun *Introduction to foliations and lie groupoids*

Global Methods for Combinatorial Isoperimetric Problems

Certain constrained combinatorial optimization problems have a natural analogue in the continuous setting of the classical isoperimetric problem. The study of so-called combinatorial isoperimetric problems exploits similarities between these two, seemingly disparate, settings. This text focuses on global methods. This means that morphisms, typically arising from symmetry or direct product decomposition, are employed to transform new problems into more restricted and easily solvable settings whilst preserving essential structure.

This book is based on Professor Harper's many years experience in teaching this subject and is ideal for graduate students entering the field. The author has increased the utility of the text for teaching by including worked examples, exercises and material about applications to computer science. Applied systematically, the global point of view can lead to surprising insights and results and established researchers will find this to be a valuable reference work on an innovative method for problem solving.

Global Methods for Combinatorial Isoperimetric Problems

L. H. HARPER

CAMBRIDGE UNIVERSITY PRESS
Cambridge, New York, Melbourne, Madrid, Cape Town, Singapore,
São Paulo, Delhi, Dubai, Tokyo, Mexico City

Cambridge University Press
The Edinburgh Building, Cambridge CB2 8RU, UK

Published in the United States of America by Cambridge University Press, New York

www.cambridge.org
Information on this title: www.cambridge.org/9780521183833

First published 2004
First paperback edition 2010

A catalogue record for this publication is available from the British Library

ISBN 978-0-521-83268-7 Hardback
ISBN 978-0-521-18383-3 Paperback

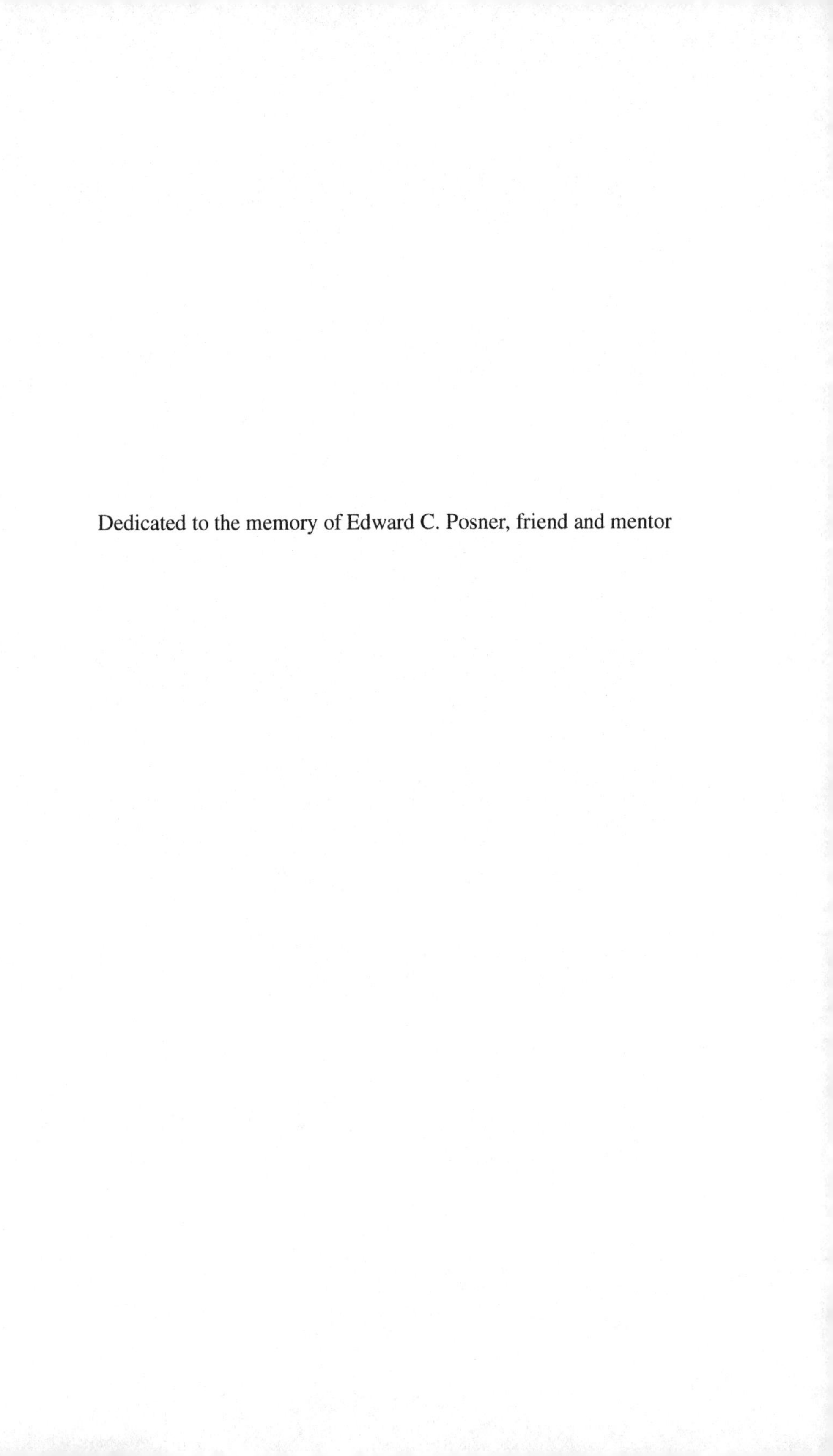

Dedicated to the memory of Edward C. Posner, friend and mentor

Contents

Preface

The purpose of this monograph is a coherent introduction to global methods in combinatorial optimization. By "global" we mean those based on morphisms, i.e. maps between instances of a problem which preserve the essential features of that problem. This approach has been systematically developed in algebra, starting with the work of Jordan in 1870 (see [**90**]). Lie's work on continuous groups, which he intended to apply to differential equations, and Klein's work on discrete groups and geometry (the Erlanger program) resulted from a trip the two made to Paris where they were exposed to Jordan's ideas. Global methods are inherent in all of mathematics, but the benefits of dealing with morphisms do not always justify the effort required and it has also been ignored in many areas. This has been especially true of combinatorics which is viewed by most of its practitioners as the study of finite mathematical structures, such as graphs, posets and designs, the focus being on problem-solving rather than theory-building.

What kinds of results can global methods lead to in combinatorics? Notions of symmetry, product decomposition and reduction abound in the combinatorial literature and these are by nature global concepts. Can we use the symmetry or product decomposition of a particular combinatorial problem to systematically reduce its size and complexity? Many of our results give positive answers to this question. We are not claiming, however, that the global point of view is the only valid one. On the contrary, we are endeavoring to show that global methods are complimentary to other approaches. Our focus is on global methods because they present opportunities which still remain largely unexploited.

The history of mathematics shows that "point of view" can be very important. What is difficult from one point of view may become easy from another. The classical Greek problems of constructing tangents for a plane curve and calculating the area enclosed by such curves were effectively solved only after the introduction of Cartesian coordinates. This allowed geometry to be translated into algebra, from which the patterns of the solutions sprang forth, creating calculus. Generally, the more varied and effective the points of view which a subject admits, the more profound and useful it becomes. It is our contention that the global point of view is effective for at least two of the most important problems of combinatorial optimization, namely the minimum path problem and the maximum flow problem (see **[6]**, or one of the many other books on algorithmic analysis, which will verify the predominance of these two problems). It has made some results easier to discover, easier to prove, easier to communicate and teach, and easier to generalize. This monograph was written to demonstrate the validity of these claims for isoperimetric problems on graphs, a subject closely related to the minimum path problem.

In this monograph, morphisms are used to represent reductions (simplifications) of a problem. Such a morphism typically maps a structure, representing an instance of a problem, to another structure of the same kind, in such a way that the essence of the problem is preserved. The latter structure, the *range* of the morphism, is typically smaller than the former, the *domain*. But since a morphism "preserves" the problem, solving it on the range will give the solution for the domain. In this volume we only deal with morphisms in an elementary way so there is no need to use (or even know) category theory. The reader should be aware, however, should questions arise, that category theory is the road map of morphism country.

The term "combinatorial (or graphical) isoperimetric problem" is now part of the language of combinatorics, but its first use, 35 years ago in the title of **[46]**, was intended to be somewhat shocking. The classical isoperimetric problem of Greek geometry is inherently continuous, involving notions of area, and length of boundary, whereas combinatorial structures are finite and inherently discontinuous. How can they go together? The apparent oxymoron was applied to graphs and posets in an effort to draw attention to an analogy between certain natural constrained optimization problems on those structures and the classical isoperimetric problem of Euclidean geometry. Initially, the nomenclature was meant to reinforce the idea that these combinatorial problems were fundamental and therefore deserving of further study. As we see in this text and others, the analogy has also shown the way for adapting powerful algebraic and analytic tools from classical mathematics to solve combinatorial problems.

There are three classes of previous publications (monographs) which relate to this one:

(1) Surveys of combinatorial optimization, including isoperimetric problems, focusing on results:
   (a) *Sperner Theory in Partially Ordered Sets* by K. Engel and H.-D. O. F. Gronau (1985) **[35]**

(b) *Combinatorics of Finite Sets* by I. Anderson (1987) **[7]**
(c) *Sperner Theory* by K. Engel (1997) **[34]**.

(2) One book which develops global methods, Steiner symmetrization and its variants, for solving (continuous) isoperimetric problems arising in applications: *Isoperimetric Inequalities in Mathematical Physics* by G. Polya and G. Szegö (1951) **[84]**.

(3) Research monographs which develop discrete analogs of harmonic and spectral analysis to solve combinatorial problems related to isoperimetry:
(a) *Group Representations in Probability and Statistics* by P. Diaconis (1988) **[33]**
(b) *Discrete Groups, Expanding Graphs and Invariant Measures* by A. Lubotzky (1994) **[75]**
(c) *Spectral Graph Theory* by F. R. K. Chung (1997) **[27]**.

This volume may be thought of as in the same spirit as the monographs of (3), taking the combinatorial isoperimetric problems of (1) and developing the discrete analogs of the global methods of (2) to solve them. The reality behind it, however, was a bit more complex. Compression was already a standard tool of combinatorialists in the mid-1970s when the author asked the questions (above, second paragraph) leading to the development of stabilization. In about 1976 G.-C. Rota, an outstanding mathematical scholar as well as one of the best listeners of the author's acquaintance, remarked on the analogy between stabilization and Steiner symmetrization, referring to the Polya–Szegö monograph. Even though the definitions of stabilization and compression are simple and natural in the combinatorial context, it seems unlikely that they would have been found by anyone deliberately seeking a discrete analog of symmetrization. Furthermore, the combinatorial setting leads naturally to the definition of partial orders, called stability and compressibility orders in this monograph, which characterize stable or compressed sets as ideals. This goes beyond the Polya–Szegö theory and it is not yet clear how to define the corresponding structures in the continuous context. However Steiner's historical model of global methods, with its wealth of applications, has given guidance and the assurance of depth to the combinatorial project.

It is probably too early yet to make a definitive statement about how global methods stack up against harmonic and spectral analysis (harmonic analysis may be identified with the spectral analysis of a Laplacian) as they are all still being developed and applied. They each have historical roots in the early nineteenth century: harmonic analysis beginning with Fourier, spectral analysis with Sturm–Liouville and global methods with Steiner (symmetrization). The author first confronted the question of how they relate in 1969: Having solved the wirelength and bandwidth problems on the d-cube where the objective functions are L_1 and L_∞ norms respectively, it was natural to pose the L_2 analog but the combinatorial methods which had been successful in L_1 and L_∞ did not go very far in L_2. Several years later a beautiful solution to the L_2 problem, using harmonic analysis on the group 2^d, was published by Crimmins, Horwitz, Palermo and Palermo [**30**]. On the other hand, harmonic analysis on the group 2^d did not seem capable of matching the results produced by global methods on the L_1 and L_∞ problems, so we concluded that the two approaches are complimentary, each majorizing the other on different problems. Recently, a superficial examination of the evidence which has accumulated since 1969 led to the same conclusion. It would seem fair to say that stabilization, the Steiner operation based on symmetry, is very limited compared to harmonic analysis. After all, it only works for reflective symmetry of a special kind. But for many of the problems where it does apply (and there are a number of important ones), the results from stabilization are more accurate and seem likely to remain so.

One of the most exciting prospects for applications of isoperimetric inequalities in recent years is the connection with the rate of convergence of a random walk, the focus of Diaconis's monograph [**33**]. After Diaconis calculated that it takes seven riffle shuffles to randomize a deck of cards, it became a legal requirement for black jack dealers in Las Vegas. The same mathematics is at the foundation of efficient random algorithms for many problems which would otherwise be intractable.

Applications are the touchstone of mathematics. The author started solving combinatorial isoperimetric problems as a research engineer in communications at the Jet Propulsion Laboratory. Since then, as a mathematician at the Rockefeller University and the University of California at Riverside, applications to science and engineering have continued to motivate the work. A good application for the solution of a hard problem doubles the pleasure, and every other benefit, from it. Global methods are by nature abstract and might easily degenerate into what von Neumann called "baroque mathematics" if not guided by real applications. On several occasions over the years, promising technical insights were left undeveloped until the right application came along. We would recommend that same caution to others developing global methods.

This monograph grew out of lectures given in the graduate combinatorics course at the University of California, Riverside from 1970 to the present (2003). The first five chapters develop the core concepts of the theory and have been pretty much the same since 1990. The development is pedestrian, assumes only an elementary knowledge of combinatorics, and largely follows the logic of discovery:

Chapter 1 is preglobal: Defining the edge-isoperimetric problem, solving it for the d-cube and presenting several applications to engineering problems.

Chapter 2 is transitional: Bringing out the connection between the edge-isoperimetric problem and the minimum path problem on networks and observing that the minimum path problem has a natural notion of morphism which extends to the edge-isoperimetric problem.

Chapter 3 is central: Defining stabilization and compression, developing their basic theory and demonstrating its efficacy in solving edge-isoperimetric problems.

Chapter 4 reinforces 1–3: Defining the vertex-isoperimetric problem and showing how stabilization and compression are also effective on it.

Chapter 5: Begins the process of deepening and extending the theory of stabilization, mainly by connecting it with Coxeter's theory of groups generated by reflections.

Those first five chapters have been used repeatedly (the course was generally offered every second or third year) for a one-quarter graduate course. The last five chapters extend the core material of Chapter 3 in five different directions. Since our thesis is that systematizing and refining the ideas that had solved challenging combinatorial isoperimetric problems would open up new possibilities, demonstrations were required to make the case. The last five chapters, which resulted from research based on the first five, constitute the necessary demonstrations.

Chapter 6: Begins the process of deepening and extending the theory of compression.

Chapter 7: Extends the theory of stabilization to infinite graphs.

Chapter 8: Extends the isoperimetric problems and their global methods to higher dimensional complexes (hypergraphs).

Chapter 9: Builds on the results of chapters 3, 4, 5 and 6 which show that isoperimetric problems on graphs can, in many interesting cases, be reduced to maximizing the weight of an ideal of fixed cardinality in a weighted poset. A new notion of morphism for this maximum weight ideal problem is introduced and applied to solve several combinatorial isoperimetric problems which seemed impossibly large.

Chapter 10: Reintroduces one of the oldest tools of optimization: calculus. The main combinatorial tool for solving isoperimetric problems on infinite families of graphs is compression. Compression requires that all members of the family have nested solutions and this is not the case for some isoperimetric problems that arise frequently in applications. Passage to a continuous limit which may then be solved by analytic methods can give asymptotic solutions, and in some cases even exact solutions. In the most interesting cases passage to the limit is facilitated by stabilization.

The last five chapters could be covered in a second quarter course. A semester course should cover the first five chapters and then a selection of the last five as time allows. Some of them presume a bit of background in algebra or analysis. The book may also be used for self-study, in which case we recommend that it be studied top-down, rather than front-to-back. If a concept does not sit well, look at its variants and analogs as they occur elsewhere in the text. That was, after all, how the material developed in the first place.

I wish to thank my students over the years for their effort, patience and helpful feedback, especially Joe Chavez who prepared a preliminary version of the manuscript. Conversations with Sergei Bezrukov, Konrad Engel and Gian-Carlo Rota have also had a profound effect on my thinking about the subject.

Working on combinatorial isoperimetric problems, from the summer of 1962 to the present, has been the greatest aesthetic experience of my life. Coxeter, in his classic monograph on regular polytopes [**28**], points out that the theory of groups generated by reflections (which underpins our concept of stabilization), is also the mathematical basis of the kaleidoscope. The word kaleidoscope is derived from Greek which translates as "beautiful thing viewer." I hope that some of those beautiful things are visible in this presentation.

1

The edge-isoperimetric problem

1.1 Basic definitions

A *graph*, $G = (V, E, \partial)$, consists of a vertex-set V, edge-set E and boundary-function $\partial : E \to \binom{V}{1} + \binom{V}{2}$ which identifies the pair of vertices (not necessarily distinct) incident to each edge. Graphs are often represented by *diagrams* where vertices are points, and edges are curves connecting the pair of incident vertices. For any graph G, and $S \subseteq V$, we define

$$\Theta(S) = \{e \in E : \partial(e) = \{v, w\},\ v \in S \,\&\, w \notin S\},$$

and call it the *edge-boundary* of S. Then given a graph G, and $k \in \mathbb{Z}^+$, *the edge-isoperimetric problem* (EIP) is to minimize $|\Theta(S)|$ over all $S \subseteq V$ such that $|S| = k$. Note that $|\Theta(S)|$ is an *invariant*, i.e. if $\phi : G \to H$ is a graph isomorphism, then $\forall S \subseteq V_G$, $|\Theta(\phi(S))| = |\Theta(S)|$. Thus subsets of vertices which are equivalent under an automorphism will have the same edge-boundary.

Loops, i.e. edges incident to just one vertex, are irrelevant to the EIP so we shall ignore them. Most, but not all, of our graphs will be *ordinary graphs*, i.e. having no loops and no more than one edge incident to any pair of vertices. The representation of an ordinary graph may be shortened to (V, E), where $E \subseteq \binom{V}{2}$, and ∂ is implicitly the identity.

1.2 Examples

1.2.1 K_n, the complete graph on n vertices

K_n has n vertices with $E = \binom{V}{2}$, i.e. there is an edge between every pair of distinct vertices. For every $S \subseteq V$ such that $|S| = k$, $|\Theta(S)| = |S \times (V - S)| = k(n - k)$. So the EIP on K_n is easy: any k-set is a solution.

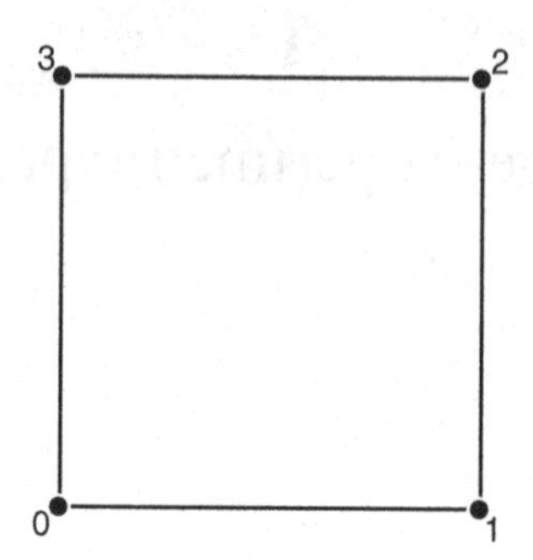

Fig. 1.1 The graph of $\mathbb{Z}_4$.

1.2.2 $\mathbb{Z}_n$, the n-cycle

For $\mathbb{Z}_n$, $V = \{0, 1, ..., n-1\}$ and $\{i, j\} \in E$ iff $i - j \equiv \pm 1 \pmod{n}$. Thus $\mathbb{Z}_3 = \mathbb{K}_3$ and $\mathbb{Z}_4$ has the diagram of Fig. 1.1.

We now deduce the solution of the EIP for $\mathbb{Z}_4$ and then $\mathbb{Z}_n$ from the following general remarks which will be useful later:

(1) (a) For $|S| = k = 0$, on any graph, there is only one subset, the empty set, $\emptyset$. Thus $\min_{|S|=0} |\Theta(S)| = |\Theta(\emptyset)| = 0$.
(b) For $k = |V| = n$, there is also only one subset, V, and $\min_{|S|=n} |\Theta(S)| = |\Theta(V)| = 0$.

(2) A graph is called *regular of degree* δ if it has exactly δ edges incident to each vertex. On a regular graph, if $|S| = k = 1$ then $|\Theta(S)| = \delta$, so any singleton is a solution set. $\mathbb{Z}_n$ is regular of degree 2; however for $n = 4$ and $k = 2$ there are two sets not equivalent under the symmetries of $\mathbb{Z}_n$: $\{0, 1\}$ and $\{0, 2\}$. All other 2-sets are equivalent to one of these two. $|\Theta(\{0, 1\})| = 2$ and $|\Theta(\{0, 2\})| = 4$, so $\min_{|S|=2} |\Theta(S)| = 2$.

(3) $\forall G$ and $\forall S \subseteq V$,

$$\Theta(V - S) = \Theta(S).$$

So for $k > \frac{1}{2}|V|$, $\min_{|S|=k} |\Theta(S)| = \min_{|S|=n-k} |\Theta(S)|$, where $n = |V|$. This completes our solution of the exterior EIP for $\mathbb{Z}_4$. It is summarized in the table

k	0	1	2	3	4
$\min_{\lvert S\rvert=k} \lvert\Theta(S)\rvert$	0	2	2	2	0

(4) Let

$$E(S) = \{e \in E : \partial(e) = \{v, w\},\ v \in S \ \&\ w \in S\}.$$

$E(S)$ is called the *induced* edges of S. The *induced edge problem* on a graph is to maximize $|E(S)|$ over all $S \subseteq V$ with $|S| = k$.

Lemma 1.1 *If $G = (V, E, \partial)$ is a regular graph of degree δ and $S \subseteq V$, then $\forall S \subseteq V$,*

$$|\Theta(S)| + 2|E(S)| = \delta |S|.$$

Proof $\delta|S|$ counts the edges incident to S but those in $E(S)$ are counted twice. □

Corollary 1.1 *If G is a regular graph, then $S \subseteq V$ is a solution of the induced edge problem iff it is a solution of the EIP. Also, $\forall k$, $\min_{|S|=k} |\Theta(S)| = \delta k - 2\max_{|S|=k} |E(S)|$.*

For regular graphs then, the EIP and induced edge problem are equivalent and we shall treat them as interchangeable. In general the EIP occurs in applications and the induced edge problem is easier to deal with in proofs. There is also a third natural variant of the EIP: for $S \subseteq V$ let

$$\partial^*(S) = \{e \in E : \partial(e) \cap S \neq \emptyset\},$$

the set of edges incident to S.

Exercise 1.1 *Show that for regular graphs, computing*

$$\min_{\substack{S \subseteq V \\ |S|=k}} \left|\partial^*(S)\right|$$

is equivalent to the EIP.

Recall that a *tree* is a graph which is connected and acyclic. An acyclic graph is also called a *forest* because it is a union of trees, its connected components.

Lemma 1.2 *The number of edges in a tree on n vertices is $n - 1$. The number of induced edges in a forest is then $n - t$, t being the number of trees.*

Any proper subset, S, of $\mathbb{Z}_n$ will induce an acyclic graph so $\max_{|S|=k} |E(S)|$ will occur for a connected set, i.e. an interval. Thus if $0 < k < n$, $\min_{|S|=k} |\Theta(S)| = 2k - 2(k-1) = 2$.

1.2.3 The d-cube, Q_d

The graph of the d-dimensional cube, Q_d, has vertex-set $\{0, 1\}^d$, the d-fold Cartesian product of $\{0, 1\}$. Thus $n = |V_{Q_d}| = 2^d$. Q_d has an edge between two vertices (d-tuples of 0s and 1s) if they differ in exactly one entry.

Exercise 1.2 *Find a formula for* $m = |E_{Q_d}|$.

Q_1 is isomorphic to K_2 and Q_2 is isomorphic to $\mathbb{Z}_4$, for which the EIP has already been solved. The 3-cube has eight vertices, 12 edges and six square faces. A diagram of Q_3., actually a projection of the 3-cube, is shown in Fig. 1.2.

One can solve the EIP on Q_3 with the simple tools which we developed in the first two examples. First observe that Q_3 has *girth* (the minimum length of any cycle) 4: since the symmetry group of the 3-cube is transitive, any vertex is the same as any other. Starting from a vertex and tracing out paths, one sees that there are no closed paths of length 3. Thus for $1 \leq k \leq 3$ we have, by Lemma 1.1 and Lemma 1.2,

$$\begin{aligned} \min_{|S|=k} |\Theta(S)| &= 3k - 2\max_{|S|=k} |E(S)| \\ &= 3k - 2(k-1) = k + 2. \end{aligned}$$

For $k = 4$ either the set induces a cycle, in which case it is a 4-cycle and has $|\Theta(S)| = 4$, or the induced graph is acyclic and by the above, $|\Theta(S)| \geq 6$. For

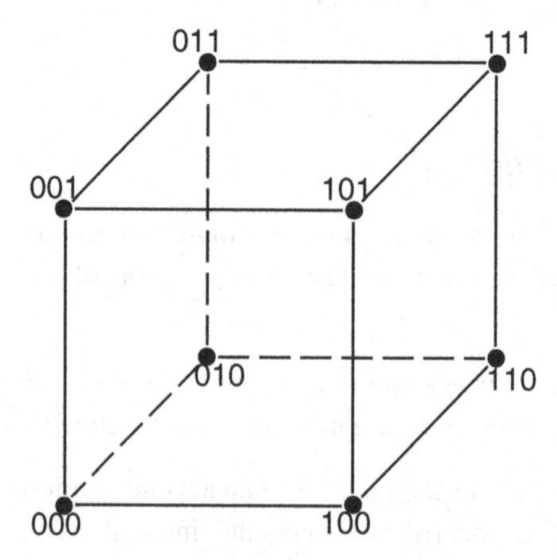

Fig. 1.2 Diagram of Q_3.

$k > 4 = \frac{8}{2}$ we complete the solution from the fact that

$$\min_{|S|=k} |\Theta(S)| = \min_{|S|=8-k} |\Theta(S)| .$$

Our solution is summarized in the table

k	0	1	2	3	4	5	6	7	8
$\min_{\vert S\vert=k} \vert\Theta(S)\vert$	0	3	4	5	4	5	4	3	0

In order to extend this solution of the EIP to $Q_d, d > 3$, we need some simple facts about cubes which we leave as exercises. A *c-subcube of the d-cube* is the subgraph of Q_d induced by the set of all vertices having the same (fixed) values in some $d - c$ coordinates.

Exercise 1.3 *Show that any c-subcube of the d-cube is isomorphic to the c-cube.*

Exercise 1.4 *How many c-subcubes of the d-cube are there?*

A *neighbor* of a c-subcube of the d-cube is any c-subcube which differs from the given one in exactly one of their $d - c$ fixed coordinates.

Exercise 1.5 *Show that all neighbors of a c-subcube are disjoint.*

Exercise 1.6 *Show that no (vertices in) two distinct neighbors of a c-subcube are connected by an edge.*

Exercise 1.7 *How many neighbors does a c-subcube of the d-cube have?*

The EIP on Q_d was originally motivated by problems in data transmission (see Chapter 2). Studies by W. H. Kautz [**59**], E. C. Posner (personal communication) and the author led to the conjecture that the initial segments of the *lexicographic numbering*,

$$lex(x) = 1 + \sum_{i=1}^{d} x_i 2^{i-1}$$

for $x \in V_{Q_d}$, were solution sets, but how was this to be proven? An obvious approach to try was induction on the dimension, d. Mathematical induction has the paradoxical property that it is often easier to prove a stronger theorem because, once the initial case has been verified, one is allowed to assume that the theorem is true for lower values of the inductive parameter in order to establish it for the next one. Thus a stronger hypothesis can produce an easier proof. In this case the strategy led to the conjecture that the following inductive procedure would produce *all* solution sets:

(1) Begin with the null set, $\emptyset$.
(2) Having constructed a set $S \subset V_{Q_d}$, augment it with any $x \in V_{Q_d} - S$ which maximizes the marginal number of induced edges,

$$|E(S + \{x\})| - |E(S)|.$$

Thus the augmentation of $\emptyset$ may be by any $x \in V_{Q_d}$ since $|E(\{x\})| - |E(\emptyset)| = 0$. The augmentation of $\{x\}$ must be by y which is a neighbor of x, etc. What kinds of k-sets are these for $k > 2$? The answer follows from the fact that if $k = 2^c$, then the set must be a c-subcube. We have just verified this for $c = 0, 1$. Assume it is true for $0, 1, ..., c - 1$. In augmenting a 2^{c-1}-set, which must be a $(c - 1)$-subcube, we may only choose a vertex whose marginal contribution to $|E(S)|$ is 1, i.e. any member of a neighboring $(c - 1)$-subcube. Having chosen a vertex from one of those neighboring subcubes we must continue to choose vertices from the same subcube until it is exhausted, since there will always be a vertex in the chosen subcube for which $|E(S + \{x\})| - |E(S)| \geq 2$ whereas any vertex not in the subcube will have $|E(S + \{x\})| - |E(S)| \leq 1$. When we exhaust the neighboring $(c - 1)$-subcube, we have completed a c-subcube.

In general, let

$$k = \sum_{i=1}^{K} 2^{c_i}, \quad 0 \leq c_1 < c_2 < ... < c_K,$$

be the base 2 representation of k (note that $K = \lfloor \log_2 k \rfloor$). If $S \subseteq V_{Q_d}$ is a disjoint union of c_i-subcubes, $1 \leq i \leq K$, such that each c_i-subcube lies in a neighbor of every c_j-subcube for $j > i$, then S is called *cubal*. The cubal sets are exactly the sets constructed by successively maximizing the marginal number of interior edges. It follows that if S is cubal and $|S| = k$ then (see Exercise 1.1)

$$|E(S)| = \sum_{i=1}^{K} (K - i) 2^{c_i} + c_i 2^{c_i - 1}.$$

Note that $|E(S)|$ for a k-cubal set, $S \subseteq V_{Q_d}$, does not depend on d, just on $k = |S|$. This function is important so we denote it by $E(k)$. $E(k)$ has a fractal nature which is hinted at by the following recurrence. If $2^{d-1} \leq k < 2^d$ then

$$E(k + 1) - E(k) = E\left(k - 2^{d-1} + 1\right) - E\left(k - 2^{d-1}\right) + 1.$$

This follows from the recursive structure of k-cubal sets. Subtracting the largest power of 2 from k, 2^{d-1}, corresponds to removing the largest subcube from S. That subcube provided one neighbor to every vertex in the remainder of the set.

Exercise 1.8 *Show that if $S \subseteq V_{Q_d}$ is cubal, then its complement $V_{Q_d} - S$ is cubal.*

Exercise 1.9 *Show that any two cubal k-sets are isomorphic, i.e. there is an automorphism of the d-cube which takes one to the other.*

Theorem 1.1 *$S \subseteq V_{Q_d}$ maximizes $|E(S)|$ for its cardinality, k, iff S is cubal.*

Lemma 1.3 *(Bernstein, [13]) $\forall d$ and $\forall k, t > 0$ such that $k + t < 2^d$,*

$$E(t) < E(k+t) - E(k) < E\left(2^d\right) - E\left(2^d - t\right).$$

Proof (of the lemma). By induction on d: It is true for $d = 2$. Assume it for $d - 1 \geq 2$ and consider the following three cases:

(1) If $k \geq 2^{d-1}$ then

$$\begin{aligned} E(k+t) - E(k) &= \sum_{i=1}^{t} E(k+i) - E(k+i-1) \\ &= \sum_{i=1}^{t} \left[E\left(k+i-2^{d-1}\right) - E\left(k+i-2^{d-1}-1\right) + 1\right] \\ &= E\left(k+t-2^{d-1}\right) - E\left(k-2^{d-1}\right) + t, \end{aligned}$$

and both inequalities follow from the inductive hypothesis.

(2) If $k + t \leq 2^{d-1}$ then the lefthand inequality follows from the inductive hypothesis and the righthand one from the above identity and then the inductive hypothesis.

(3) If $k < 2^{d-1} < k + t$ then

$$\begin{aligned} E(k+t) - E(k) &= \left[E(k+t) - E\left(2^{d-1}\right)\right] + \left[E\left(2^{d-1}\right) - E(k)\right] \\ &> \left[E\left(k+t-2^{d-1}\right)\right] + \left[E(t) - E\left(t - \left(2^{d-1} - k\right)\right)\right] \\ &\qquad \text{by Case 1 and Case 2, respectively,} \\ &= E(t). \end{aligned}$$

Exercise 1.10 *Complete the proof (the righthand inequality) of Case 3.*

□

Proof (of the theorem). We have noted that all k-cubal sets have the same number, $E(k)$, of induced edges, so we need only show that all optimal sets are cubal. We proceed by induction on d. We have already shown it to be true for $d = 1, 2$. Assume that it is true for all dimensions less than $d > 2$. Given k, $0 < k < n = 2^d$, with the representation as a sum of powers of 2 above (so

$K < d$), let $S \subseteq V_{Q_d}$ be optimal for $|S| = k$. If we divide Q_d into two $(d-1)$-subcubes, $Q_{d,0} = \{x \in V_{Q_d} : x_d = 0\}$ and $Q_{d,1}$, by its dth coordinate, then we partition S into $S_0 = S \cap Q_{d,0}$ and $S_1 = S \cap Q_{d,1}$. Letting $|S_0| = k_0$ and $|S_1| = k_1$, we may assume that $k_0 \geq k_1$. If $k_1 = 0$ the theorem follows from the inductive hypothesis, so assume $k_1 > 0$. The edges of $E(S)$ will either have both ends in S_0, both ends in S_1 or one end in S_0 and one in S_1. Therefore

$$|E(S)| \leq \max_{|S|=k_0} |E(S)| + \max_{|S|=k_1} |E(S)| + k_1.$$

If we take S_1 to be a cubal set of cardinality k_1, by induction, $|E(S_1)| = E(k_1) = \max_{|S|=k_1} |E(S)|$. The neighbors of S_1 in $Q_{d,0}$ are isomorphic to S_1 and so are also cubal. By the recursive construction of cubal sets, $\exists S_0 \subseteq Q_{d,0}$, cubal with $|S_0| = k_0$ and containing the neighbors of S_1. Thus the upper bound can be achieved and every set which achieves it must be such a union of two cubal sets. Let $k_0 = \sum_{i=1}^{K_0} 2^{c_{i,0}}$, $0 \leq c_{1,0} < c_{2,0} < \ldots < c_{K_o,0}$ and similarly for k_1. Since $k_0 + k_1 = k$ there are just three possibilities:

(1) $c_{K_0,0} = c_K$: Then we may assume that $S_0 - Q_{c_{K_0},0}$ and S_1 are in two distinct neighbors of $Q_{c_{K_0},0}$ so S is not cubal. By Exercise 1.5 there can be no edges between a member of $S_0 - Q_{c_{K_0},0}$ and a member of S_1. With $k_0' = k_0 - 2^{c_{l_0}} > 0$ we have $k_0' + k_1 \leq 2^{c_{K_0}}$. If we alter S by removing S_1 and adding the same number of vertices to S_0, Lemma 1.3 shows that $|E(S)|$ will be increased. This contradicts the optimality of S.

(2) $c_{K_0,0} = c_K - 1$ and $c_{K_1,1} = c_K - 1$: The $(c_K - 1)$-subcubes, $Q_{c_{K_0},0}$ and $Q_{c_{K_1},1}$, are neighbors and so constitute a c_K-subcube. $S_0 - Q_{c_{K_0},0}$ and $S_1 - Q_{c_{K_1},1}$ each lie in neighboring $(c_K - 1)$-subcubes which together constitute a c_K-subcube neighboring the first. By the inductive hypothesis S must be cubal.

(3) $c_{K_0,0} = c_K - 1$ and $c_{K_1,1} < c_K - 1$: As in Case 1, we assume that $S_0 - Q_{c_{K_0},0}$ and S_1 are in two distinct neighbors of $Q_{c_{K_0},0}$ and so have no edges between them. Not only is $k_0' = k_0 - 2^{c_{K_0}} > 0$ but

$$\begin{aligned} k_0' + k_1 &= k_0 - 2^{c_{K_0}} + k_1 \\ &= k - 2^{c_K - 1} \\ &\geq 2^{c_K} - 2^{c_K - 1} \\ &= 2^{c_K - 1}. \end{aligned}$$

If we alter S by removing $2^{c_{K_0}} - k_0'$ members of S_1 and using them to complete the neighbor of $Q_{c_{K_0},0}$ which contains $S_0 - Q_{c_{K_0},0}$, Lemma 1.3 shows that this will increase $|E(S)|$, again contradicting the optimality of S.

□

1.3 Application to layout problems

Combinatorial isoperimetric problems arise frequently in communications engineering, computer science, the physical sciences and mathematics itself. We do not wish to cover all applications here but to give a representative sample. We have chosen to concentrate on layout problems. These arise in electrical engineering when one takes the wiring diagram for some electrical circuit and "lays it out" on a chassis, i.e. places each component and wire on the chassis. A wiring diagram is essentially a graph, the electrical components being the vertices and the wires connecting them being the edges. For any given placement of the vertices and edges on the chassis, there will be certain costs or measurements of performance which we wish to optimize.

1.3.1 The wirelength problem

Suppose that we wish to place components (vertices of the graph $G = (V, E, \partial)$) on a linear chassis, each a unit distance from the preceding one, in such a way as to minimize the total length of all the wires connecting them. To make this precise, we define a *vertex-numbering* of G to be a function

$$\eta : V \to \{1, 2, ..., n\}, \text{ where } n = |V|,$$

which is one-to-one (and therefore onto). The integers in the range of η may be identified with positions on the linear chassis. The *(total) wirelength of* η is then

$$wl(\eta) = \sum_{\substack{e \in E \\ \partial(e) = \{v, w\}}} |\eta(v) - \eta(w)| .$$

For a graph, $G = (V, E, \partial)$,

$$wl(G) = \min \{wl(\eta) : \eta \text{ is a vertex-numbering of } G\} .$$

Remember that a graph on n vertices has $n!$ vertex-numberings.

1.3.1.1 Example

The graph of the square has $4! = 24$ vertex-numberings, but it also has eight symmetries. Any two numberings symmetric to each other have the same wirelength. The three numberings in Fig. 1.3 are representative of the $24/8 = 3$ equivalence classes of numberings. Thus the first two numberings minimize the wirelength, wl, and the third maximizes it. Therefore $wl(Q_2) = 6$. $wl(Q_3)$ is not so easily determined since Q_3 has $8! = 40\,320$

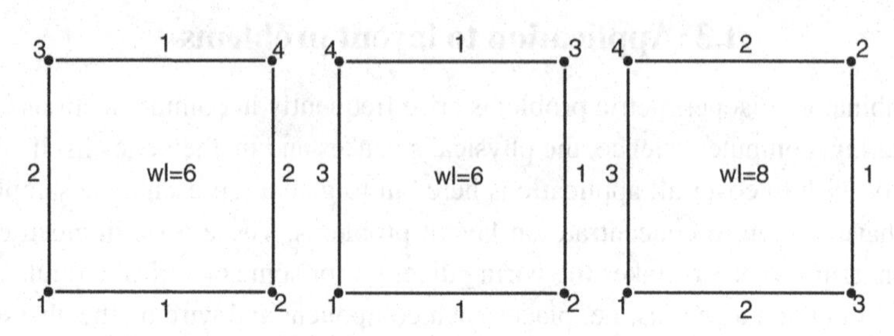

Fig. 1.3 Numberings of Q_2.

numberings and 48 symmetries. $40\,320/48 = 840$ is not all that large, but how does one systematically generate representatives of those 840 equivalence classes of numberings? We now show how to get around these apparent difficulties.

In seeking to minimize a sum like wl, an obvious strategy is to minimize each summand separately. The sum of those minima is a lower bound on the minimum of the sum and one could hope that it would be a good lower bound, even sharp. That is not the case for the definition of $wl(\eta)$, however, since for every edge $e \in E$ with $\partial(e) = \{v, w\}$

$$\min_{\eta} |\eta(v) - \eta(w)| = 1.$$

1.3.1.2 Another representation of wl

Given a numbering, η, and an integer k, $0 \leq k \leq n$, let

$$S_k(\eta) = \eta^{-1}(\{1, 2, ..., k\}) = \{v \in V : \eta(v) \leq k\},$$

the set of the first k vertices numbered by η. Then we have the following alternative representation of the wirelength.

Lemma 1.4

$$wl(\eta) = \sum_{k=0}^{n} |\Theta(S_k(\eta))|$$

Proof Note that $S_0(\eta) = \eta^{-1}(\emptyset) = \emptyset$. Let

$$\chi(e, k) = \begin{cases} 1 & \text{if } \partial(e) = \{v, w\}, \eta(v) \leq k < \eta(w) \\ 0 & \text{otherwise.} \end{cases}$$

Then

$$wl(\eta) = \sum_{\substack{e \in E \\ \partial(e) = \{v,w\}}} |\eta(v) - \eta(w)| = \sum_{e \in E} \sum_{k=0}^{n} \chi(e,k)$$

$$= \sum_{k=0}^{n} \sum_{e \in E} \chi(e,k) = \sum_{k=0}^{n} |\Theta(S_k(\eta))| .$$

□

Exercise 1.11 *(Steiglitz–Bernstein,* **[87]***) Suppose that the linear chassis is actually the real line with specified sites $s_1 < s_2 < \ldots < s_n$ for the placement of components. Show then that the layout, η, which places $v \in V$ at site $s_{\eta(v)}$ will give total wirelength*

$$wl(\eta) = \sum_{k=0}^{n} (s_{k+1} - s_k) |\Theta(S_k(\eta))| .$$

The new representation of wirelength as a sum gives us another lower bound on it:

Corollary 1.2 *For any graph, G,*

$$wl(G) = \min_{\eta} wl(\eta) = \min_{\eta} \sum_{k=0}^{n} |\Theta(S_k(\eta))|$$

$$\geq \sum_{k=0}^{n} \min_{\eta} |\Theta(S_k(\eta))| = \sum_{k=0}^{n} \min_{|S|=k} |\Theta(S)| .$$

Theorem 1.2 *Any vertex-numbering, η_0, of $G = (V, E, \partial)$, whose initial segments, $S_k(\eta_0)$, $0 \leq k \leq n$, are solutions of the EIP on G, is a solution of the wirelength problem on G.*

Corollary 1.3 *The numbering of $\mathbb{Z}_n$ defined by $\eta_0(i) = 1 + i$ is a solution of the wirelength problem on $\mathbb{Z}_n$ and*

$$wl(\mathbb{Z}_n) = \sum_{k=0}^{n} \min_{|S|=k} |\Theta(S)|$$

$$= \sum_{k=1}^{n-1} 2 = 2(n-1).$$

Definition 1.1 *If $\{T_i\}_{i=1}^n$ is a sequence of totally ordered sets, then the lexicographic order on their product, $T_1 \times T_2 \times ... \times T_n$, is the total order defined by $x < y$ if $\exists m$ such that $x_1 = y_1, x_2 = y_2, ..., x_{m-1} = y_{m-1}$ and $x_m < y_m$. Any total order on a set gives a numbering, the first (least) member of the set being numbered 1, the second 2, etc. The numbering given by lexicographic order on d-tuples of 0s and 1s (vertices of Q_d) is $lex(x) = 1 + \sum_{i=1}^d x_i 2$.*

Corollary 1.4 *The lexicographic numbering of Q_d is a solution of the wirelength problem on Q_d.*

Exercise 1.12 *Show that $wl(Q_d) = 2^{d-1}(2^d - 1)$.*

Exercise 1.13 *Show that lex also solves the wirelength problem on Q_d for the more general linear chassis of Exercise 1.11.*

1.3.2 The deBruijn graph of order 4

The deBruijn graph is a *directed graph* (*digraph*) which means that there are two boundary functionals, $\partial_\pm : E \to V$, which identify the head ($\partial_+(e)$) and tail ($\partial_-(e)$) end of each edge. Diagrams for digraphs usually have an arrow on each edge pointing to its head. DB_d, *the deBruijn graph of order d*, has the same vertex-set as Q_d (the set of all d-tuples of 0s and 1s), but its edge-set is very different. E_{DB_d} is the set of all $(d+1)$-tuples of 0s and 1s and

$$\partial_-(x_1, x_2, ..., x_d, x_{d+1}) = (x_1, x_2, ..., x_d),$$
$$\partial_+(x_1, x_2, ..., x_d, x_{d+1}) = (x_2, ..., x_d, x_{d+1}).$$

For further information about deBruijn graphs we refer the reader to [**41**]. A diagram of DB_4 is shown in Fig. 1.4.

The representation of wirelength for a numbering, η, as a sum of edge-boundaries, and its application to solve the wirelength problems on $\mathbb{Z}_n$ and Q_d suggests a heuristic for minimizing it on any graph: Number the vertices $1, 2, ..., k-1, k, ..., n$ so as to minimize

$$|\Theta(S_k(\eta))| - |\Theta(S_{k-1}(\eta))|,$$

the marginal contribution of each additional vertex. Applying this heuristic to $\underline{DB_4}$ (DB_4 with its loops removed and edges undirected) we produce the numbering, η_0 of Fig. 1.5.

The following table lists the values of $|\Theta(S_k(\eta_0))|$ for k between 0 and 16:

k	0	1	2	3	4	5	6	7	8	9	10	11	12	13	14	15	16
$\lvert\Theta(S_k(\eta_0))\rvert$	0	2	4	4	6	6	6	6	6	6	8	6	6	4	4	2	0

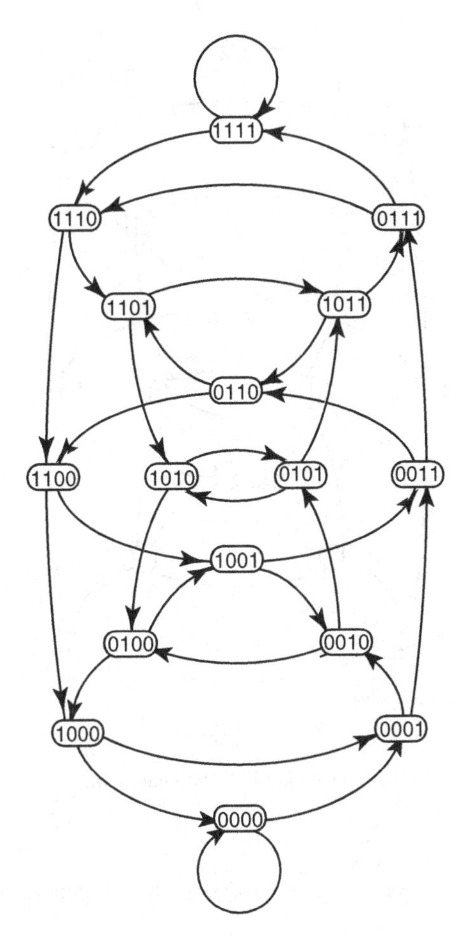

Fig. 1.4 The deBruijn graph of order 4.

From this one can see that $S_{10}\left(\eta_0\right)$ is not a solution of the EIP. In fact there is no numbering whose initial segments are all solutions of the EIP, since if there were, the procedure that we used to construct η_0 would have produced it. We claim, however, that for $k \neq 10$, $S_k\left(\eta_0\right)$ is a solution of the EIP (we will return to this claim in Chapter 5). Thus the inequality leading to Theorem 1.2 gives

$$wl\left(\underline{DB_4}\right) > \sum_{k=0}^{n} \min_{|S|=k} |\Theta(S)| = 2(34) + 6 = 74,$$

whereas

$$wl\left(\eta_0\right) = 76.$$

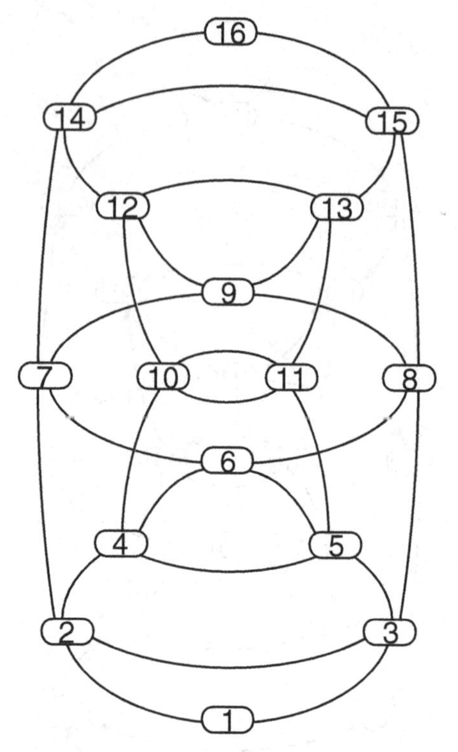

Fig. 1.5 Diagram of $\underline{DB}_4$ with numbering η_0.

Since the degree, $\delta(v)$, of each $v \in V_{\underline{DB_4}}$ is even and (generalizing the identity of Lemma 1.1)

$$|\Theta(S)| = \sum_{v \in S} \delta(v) - 2|E(S)|,$$

$|\Theta(S)|$ is even for all $S \subseteq V_{\underline{DB_4}}$, we have $wl(\eta) \geq 76$ for any numbering η of $\underline{DB_4}$ and η_0 is a solution of its wirelength problem.

Exercise 1.14 *Find* all *solution sets for the EIP on* $\underline{DB_4}$ *with* $k \leq 8$ *(proof not required, just a list).*

1.3.3 Partitioning problems

A *partition* of $G = (V, E, \partial)$ is a set $\pi \subseteq 2^V$ such that

(1) $\forall B \in \pi, B \neq \emptyset$,

(2) $\forall A, B \in \pi$ either $A = B$ or $A \cap B = \emptyset$,
(3) $\bigcup_{B \in \pi} B = V$.

$B \in \pi$ is called a *block* of the partition. A partition, π, is called *uniform* if $\forall A, B \in \pi$, $||A| - |B|| \leq 1$. If $|\pi| = p$, this is equivalent to

$$\forall B \in \pi, \quad \left\lfloor \frac{n}{p} \right\rfloor \leq |B| \leq \left\lceil \frac{n}{p} \right\rceil, \qquad \text{where } n = |V|.$$

In fact, if j is the remainder, or residue, of n divided by p (i.e. $n \equiv j \pmod{p}$, $0 \leq j < p$), then any uniform p-partition of G will have $j \left\lceil \frac{n}{p} \right\rceil$-blocks and $(p - j) \left\lfloor \frac{n}{p} \right\rfloor$-blocks.

The *edge-boundary* of a partition, π, is

$$\begin{aligned} \Theta(\pi) &= \{e \in E : \partial(e) = \{v, w\}, v \in A, w \in B \ \& \ A \neq B\} \\ &= \bigcup_{B \in \pi} \Theta(B). \end{aligned}$$

Note that the union in the latter representation is not a disjoint union, but that any $e \in E$ with $\partial(e) = \{v, w\}$ is included exactly twice, once for the block containing v and once for the block containing w.

The *edge-boundary partition problem* for G is to minimize $|\Theta(\pi)|$ over all uniform p-partitions, π, of G. One may think of G as a wiring diagram which is to be layed out on p chips, the components to be divided up among the chips as equally as possible. The problem is to assign components to chips so that the number of wires connecting components on different chips is minimized. A variant of the edge-boundary partition problem is the *edge-width partition problem* which is to minimize $\max_{B \in \pi} |\Theta(B)|$ over all uniform p-partitions, π, of G.

Lemma 1.5 *If* $n \equiv j \pmod{p}$, $0 \leq j < p$, *then*

$$\min_{\substack{|\pi| = p \\ \pi \text{ uniform}}} |\Theta(\pi)| \geq \frac{1}{2} \left(j \min_{|B| = \left\lceil \frac{n}{p} \right\rceil} |\Theta(B)| + (p - j) \min_{|B| = \left\lfloor \frac{n}{p} \right\rfloor} |\Theta(B)| \right).$$

Also,

$$\min_{|\pi| = p} \max_{B \in \pi} |\Theta(B)| \geq \max \left\{ \min_{|B| = \left\lceil \frac{n}{p} \right\rceil} |\Theta(B)|, \min_{|B| = \left\lfloor \frac{n}{p} \right\rfloor} |\Theta(B)| \right\}.$$

Proof If π_0 is an optimal uniform p-partition, then

$$\min_{|\pi|=p} |\Theta(\pi)| = |\Theta(\pi_0)|$$
$$= \frac{1}{2} \sum_{B \in \pi_0} |\Theta(B)|,$$

since each $e \in \Theta(\pi_0)$ will be counted twice in the sum,

$$\geq \frac{1}{2} \left(j \min_{|B|=\lceil \frac{n}{p} \rceil} |\Theta(B)| + (p-j) \min_{|B|=\lfloor \frac{n}{p} \rfloor} |\Theta(B)| \right).$$

□

1.3.3.1 Examples

(1) For $\mathbb{Z}_n$ we get

$$\min_{|\pi|=p} |\Theta(\pi)| \geq \frac{1}{2} \left(j \min_{|B|=\lceil \frac{n}{p} \rceil} |\Theta(B)| + (p-j) \min_{|B|=\lfloor \frac{n}{p} \rfloor} |\Theta(B)| \right)$$
$$= \frac{1}{2} (j \cdot 2 + (p-j) \cdot 2) = p.$$

Also,

$$\min_{|\pi|=p} \max_{B \in \pi} |\Theta(B)| \geq \max \left\{ \min_{|B|=\lceil \frac{n}{p} \rceil} |\Theta(B)|, \quad \min_{|B|=\lfloor \frac{n}{p} \rfloor} |\Theta(B)| \right\}$$
$$= \max\{2, \ 2\} = 2.$$

These lower bounds can be achieved by uniform p-partitions into intervals, so the edge-boundary and edge-width partition problems have been solved for $\mathbb{Z}_n$.

(2) For Q_d and $p = 2^a$, a power of 2, $\lceil \frac{n}{p} \rceil = \frac{2^d}{2^a} = 2^{d-a} = \lfloor \frac{n}{p} \rfloor$, so we get

$$\min_{|\pi|=p} |\Theta(\pi)| \geq \frac{1}{2} \left(j \min_{|B|=\lceil \frac{n}{p} \rceil} |\Theta(B)| + (p-j) \min_{|B|=\lfloor \frac{n}{p} \rfloor} |\Theta(B)| \right)$$
$$= \frac{1}{2} 2^a \min_{|B|=2^{d-a}} |\Theta(B)| = \frac{1}{2} 2^a \cdot (d-a) 2^{d-a}$$
$$= (d-a) 2^{d-1}.$$

Again, not surprisingly, this lower bound, and the corresponding one for edge-width, can be achieved by partitioning Q_d into $(d-a)$-subcubes. What is surprising, though, is the observation by Bezrukov [**15**] that the lower bounds continue to be sharp for other (and maybe all) values of p.

Theorem 1.3 *$\forall d > a$ there exists a uniform (2^a+1)-partition of Q_d into cubal sets.*

Exercise 1.15 *Before reading the following proof, prove the special case $a = 1$.*

Proof We begin with a cubal $\left\lfloor \frac{2^d}{2^a+1} \right\rfloor$-set, B_1, in Q_d. Since all cubal k-sets are equivalent under isomorphism (see Exercise 1.9), we might as well take $B_1 = S_k(lex)$, for $k = \left\lfloor \frac{2^d}{2^a+1} \right\rfloor$. Since $\left\lfloor \frac{2^d}{2^a+1} \right\rfloor < 2^{d-a}$, B_1 is a subset of the $(d-a)$-subcube of Q_d whose fixed coordinates are the last a which are all 0. The 2^a values of these fixed positions each give a copy of B_1 which is therefore cubal. Number these $B_1, B_2, \ldots, B_{2^a}$ in lex order on the fixed positions. Also add the element $S_{k+1}(lex) - S_k(lex)$ to B_1 and the corresponding element to $B_2, \ldots, B_j$, j the remainder of 2^d divided by (2^a+1), so these first j blocks are copies of $S_{k+1}(lex)$. We claim that

$$\bigcup_{i=1}^{2^{d-a}} B_i$$

is a cubal set: It consists of a union of subcubes of dimension $a_i + a$, one for each exponent in $\sum 2^{a_i} = k$ and each $(a_i + a)$-subcube is in a neighbor of all larger ones. Also it has a union of b_l-subcubes for each exponent in $\sum 2^{b_l} = j$, $b_1 < b_2 < \ldots$. Since the b_l-subcubes are all in an a-subcube which is in a neighbor of all the $(a_i + a)$-subcubes, the claim is proven. Therefore $B_{2^a+1} = V_{Q_d} - \bigcup_{i=1}^{2^{d-a}} B_i$ is a cubal set of size $\left\lfloor \frac{2^d}{2^a+1} \right\rfloor$ and by Exercise 1.8

$$\pi = \{B_1, B_2, \ldots, B_{2^a}, B_{2^a+1}\}$$

is the required partition. □

Lemma 1.6 *If Q_d has a uniform p-partition into cubal sets, then Q_{d+1} has a uniform $2p$-partition into cubal sets.*

Proof Exercise. □

Bezrukov [**15**] goes on to show by similar methods that for fixed p, Q_d has a uniform p-partition into asymptotically optimal sets as $d \to \infty$.

1.4 Comments

Ah, but a man's reach should exceed his grasp – else what's a heaven for? (Robert Browning)

I was just a beginning graduate student when I wrote the paper [**45**] upon which this chapter is based. My reach did certainly exceed my grasp, maybe even more than Browning would have approved. Fortunately, I did not have to wait for heaven to make up for my shortcomings. Two years after the publication of my paper, Bernstein published a follow-up [**13**], pointing out that I had overlooked a case (Case 3 in the argument for Theorem 1.1) and filling the gap. At the time Bernstein's patch (Lemma 1.3) seemed disappointingly complicated. Now, after almost forty years of experience with this material, I see that Bernstein's lemma was precisely what was needed. It not only covers the missing Case 3, but replaces my fuzzy argument for Case 1 with a clear one. It also contains the glimmers of deeper insights which developed later (cf. Chapter 9).

The problem of finding an optimal uniform 2-partition for the vertices of a graph is an appealing special case of the EIP with many different names, such as "graph bisection" or "minimum balanced cut" in the literature. See the Comments in Chapter 6 for more about it.

What we are calling the wirelength problem has occurred in many different applications and so goes by different names in the literature. We have settled on "wirelength" because it is succinct and descriptive. A solution of the wirelength problem for $\mathbb{Z}_n$ was published by Lehman [**72**] in 1963. Its application was to show how to construct a bracelet from n beads, of various weights but connected by uniform springs, so as to minimize the fundamental frequency. The original application for the solution of the wirelength problem for Q_d [**45**] was to encode data from the set $\{1, 2, ..., 2^d\}$ into d-tuples of 0s and 1s, so that when transmitted over a noisy (binary symmetric) channel with a low probability of error in any bit, the average absolute error would be minimized. The original instance of a wirelength problem, as such, was the deBruijn graph of order 4 [**47**] which occurred as the wiring diagram of a decoding circuit. Minimizing wirelength meant minimizing self-inductance.

The wirelength problem for Q_d, which appears in this chapter as an application, was actually the starting point for my work. Ed Posner, my boss at the Jet Propulsion Laboratory (JPL), posed the problem to me. The application he had in mind was not minimizing wirelength but minimizing the average absolute error in transmitting linear data over a binary symmetric channel. At that time JPL had video cameras which were transmitting the first close-up pictures of the surface of the Moon. The pixels in the pictures were registered as shades of grey, 64 of them, ranging from white to black. They were encoded as 6-tuples of 0s and 1s for transmission to Earth where they were decoded back into shades of grey and reassembled into a picture. The problem was that the transmitter was minimally powered and in transmission a 0 would occasionally be changed into a 1 or vice versa. This would cause an error in the resulting shade of grey (depending on the code) and degrade the received picture. The challenge was to prove that the code which the engineers were using, the lexicographic code (numbering), was one which minimized the average absolute error.

As an undergraduate, I had been impressed with the writings of G. Polya on problem-solving [**83**]. Polya's thesis is that problem-solving can be learned and in particular that there are effective strategies which can be consciously applied. One of these is that of reducing a conjecture, A, to the conjunction of, presumably easier, conjectures B and C. I followed Polya's advice in working on Posner's problem, which led to the train of thought in Section 1.3.1. When I saw that Posner's conjecture reduced to the conjecture that the initial segments of lex numbering on Q_d solved the EIP, and that *that* conjecture seemed amenable to induction on d, I was pretty confident that I was onto something. Having found it so helpful myself, I have recommended Polya's method to young mathematicians ever since.

Being able to give *all* solutions of the EIP for Q_d was a fortuitous mistake. At the time I thought that it was necessary for the logic of the proof by induction. In retrospect that is clearly not the case, but it did lead to a stronger result which has made the result more flexible to apply. Note that Bezrukov's application to edge-boundary partition problems, which utilizes that flexibility, appeared 33 years after [**45**].

2
The minimum path problem

2.1 Introduction

In this section and the next, we present the essentials of the theory of minimum path problems. The minimum path problem is arguably the most fundamental and most frequently applied of all optimization problems. These first two sections constitute background for our introduction of global methods. Readers already familiar with the material may wish to use them for a quick review and reference. The main line of our development resumes in Section 2.3 with the reduction of wirelength to minimum path.

2.1.1 Basic definitions

A *network*, $N = (G; s, t; \omega)$, consists of a *directed* graph $G = (V, E, \partial_+, \partial_-)$ (where $\partial_\pm : E \to V$ give the head and tail ends, respectively, of each edge) with distinguished vertices s and t, and a *weight function* $\omega : E \to \mathbb{R}$. A *$u$-$v$ path*, P, in G is a sequence of edges, $P = (e_1, e_2, ..., e_k)$, such that $\partial_- (e_1) = x_0 = u$, $\partial_+ (e_k) = x_k = v$ and for all i, $1 \leq i < k$, $\partial_+ (e_i) = x_i = \partial_- (e_{i+1})$ (Fig. 2.1). k, the number of edges, is called the *length* P.

A u-v path of length $k \geq 1$ for which $u = v$ is called a *cycle*. If a path does not contain any cycles it is called *simple*. Let the *weight* of a path, P, be

$$\omega(P) = \sum_{i=1}^{k} \omega(e_i) .$$

If we let $\mathfrak{P}(N)$ be the set of all s-t paths in N, then the *minimum path problem* (minpath or MPP for short) is to find

$$\min_{P \in \mathfrak{P}(N)} \omega(P) ,$$

the minimum weight of any s-t path in N.

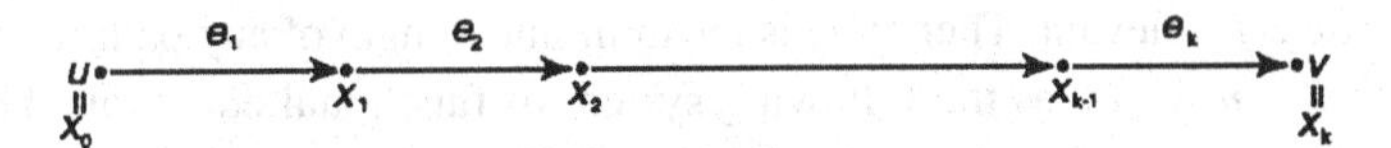

Fig. 2.1 Diagram of a u-v path of length k.

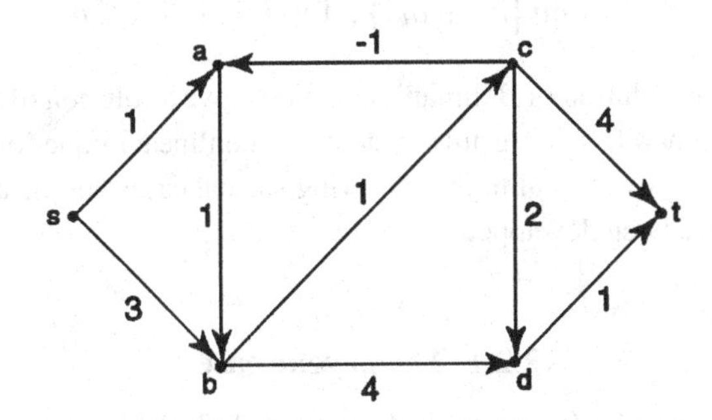

Fig. 2.2 Diagram of a network.

2.1.2 Example

In Fig. 2.2 the path $(s, b), (b, d), (d, t)$ has weight $3 + 4 + 1 = 8$, whereas the path $(s, a), (a, b), (b, d), (d, t)$ has weight $1 + 1 + 4 + 1 = 7$. Note that this network has a cycle around the vertices a, b, c of weight 1. Traversing this cycle repeatedly can create paths of arbitrarily large weight. There are, however, no negative cycles.

Exercise 2.1 *For the network of Fig. 2.1 systematically generate all simple s-t paths, evaluate their weights and find* $\min_{P \in \mathfrak{P}(N)} \omega(P)$.

2.2 Algorithms

Chapter 3 of **[68]** has a nice exposition on basic algorithms for solving the minpath problem which we summarize in this section. For a more recent and detailed exposition see **[6]**.

Let N be a network having n vertices and let $v_1, v_2, ..., v_n$ be an ordering of V such that $v_1 = s$ and $v_n = t$. Define an $n \times n$ matrix (a_{ij}) as follows:

$$a_{ij} = \min_{\substack{\partial_-(e) = v_i \\ \partial_+(e) = v_j}} \omega(e).$$

Note that if there is no edge from v_i to v_j then $a_{ij} = \infty$. This definition takes into account that for the minpath problem, only the lowest weight edge between two vertices is relevant. Then if u_j is the minimum length of any path from s to v_j, $1 \leq j \leq n$, it solves the following system of functional equations, known as *Bellman's equations* (the principle of optimality):

$$\begin{aligned} u_1 &= 0 \\ u_j &= \min_{k \neq j} \left\{ u_k + a_{kj} \right\}, \text{ for } j = 1, 2, ..., n. \end{aligned}$$

Conversely, any solution of Bellman's equations give a solution of the minpath problem. In general, solving this system of (nonlinear) equations is tricky, even if a solution exists, but in the following special cases simple and efficient algorithms have been developed.

2.2.1 The acyclic case

If a network is acyclic (contains no directed cycles), then there is an ordering of the vertices such that if $e \in E$, $\partial_- (e) = v_i$ and $\partial_+ (e) = v_j$ then $i < j$. Conversely, if there is such an ordering of V then the network is acyclic. Suppose that N is an acyclic network and that its vertices have been ordered in this manner. Then Bellman's equations reduce to:

$$\begin{aligned} u_1 &= 0 \\ u_j &= \min_{k < j} \left\{ u_k + a_{kj} \right\}, \text{ for } j = 2, ..., n. \end{aligned}$$

Note that to solve for u_j, we need only know u_k for $k < j$. This is a significant simplification in Bellman's equations. The reduced system can be solved recursively in $O(m) \leq O(n^2)$ operations, where $n = |V|$ and $m = |E|$.

2.2.2 Positive weights: Dijkstra's algorithm

If the weights on the edges in a (not necessarily acyclic) network are all positive ($\omega(e) > 0$), then the following algorithm due to Dijkstra may be used to solve Bellman's equations. The algorithm consists in putting labels on the vertices, either permanent or temporary, beginning with s as the only vertex with a permanent label and terminating when all vertices have permanent labels.

In the initial step set

$$\begin{aligned} u_1 &= 0 \\ u_j &= a_{1j}, \text{ for } j = 2, ..., n \\ P &= \{1\} \\ T &= \{2, ..., n\}. \end{aligned}$$

In the second step find $k \in T$ such that $u_k = \min_{j \in T} \{u_j\}$. Move k from T to P. If $T = \emptyset$, the process is complete. Otherwise, let

$$u_j = \min \{u_k + a_{kj}\} \text{ for all } j \in T$$

and repeat the second step.

At the conclusion of this process, the minimum weight of an s-t path is given by u_n. Dijkstra's algorithm requires $O(m) \leq O(n^2)$ operations.

2.2.3 No negative cycles: the Bellman–Ford algorithm

The Bellman–Ford algorithm uses a more complicated recursion (than either of the preceding algorithms) to solve the minpath problem on networks which contain no negative cycles. Initially we set

$$u_1^{(1)} = 0$$
$$u_j^{(1)} = a_{1j}, \text{ for } j = 2, ..., n.$$

Subsequently we let

$$u_j^{(p+1)} = \min \left\{ u_j^{(p)}, \min_{k \neq j} \left\{ u_k^{(p)} + a_{kj} \right\} \right\}.$$

$u_j^{(p)}$ may be thought of as the minimum weight of any path from s to v_j which contains no more than p edges. Thus the minimum weight of any s-t path is given by $u_n^{(n-1)}$ if N contains no negative cycles reachable by an s-t path. The algorithm may even be used to detect such negative cycles since that is the case iff $u_n^{(2n-1)} = u_n^{(n-1)}$. The Bellman–Ford algorithm requires no more than $O(m \cdot n) \leq O(n^3)$ operations.

2.2.4 The general case

If N contains no negative cycles reachable by an s-t path, then a minimum weight path must be simple and a solution can be found in $O(n^3)$ operations. If there are negative cycles reachable by an s-t path, that can also be determined in $O(n^3)$ operations and there is no minimum weight path. We might attempt to extend the domain of solvability of the minpath problem by restricting $\mathfrak{P}$ to simple paths. However, that problem is essentially the Hamiltonian path problem, known to be NP-complete and thus unsolvable in any practical sense (see p. 797 of [**6**]).

2.2.5 An observation by Klee

Note that all of these algorithms work equally well if the weight of a path, P, is

$$\mu\omega(P) = \max_{e \in P} \omega(e).$$

All we need do is replace $\{u_k + a_{kj}\}$ by $\max\{u_k, a_{kj}\}$. This is a special case of a general observation made by V. Klee. See problem 5.6 in [**68**].

2.3 Reduction of wirelength to minpath

Given a graph $G = (V, E, \partial)$, representing an instance of the wirelength problem, we wish to produce a network, $N(G)$, representing an instance of the minpath problem and such that a solution of the minpath problem on $N(G)$ gives a solution of the original wirelength problem and vice versa. There is a natural way to do this already inherent in Chapter 1. It does require, however, that we change the standard definition of a network just a bit. The weights will be on the vertices ($\omega : V \to \mathbb{R}$) with the weight of a path being $\sum_{i=0}^{k} \omega(x_i)$, the sum of the weights of the vertices in that path (see Fig. 2.1).

Exercise 2.2 *Verify that the algorithm for solving the minpath problem on an acyclic network works equally well on vertex-weighted networks.*

Given G, $N(G)$, its *derived network*, consists of a digraph $G' = (V', E', \partial_+, \partial_-)$, distinguished vertices, s, t, and a weight function (on the vertices). We take $V' = 2^V$, the set of all subsets of V, and $s = \emptyset$, $t = V$. Also

$$E' = \{(S, T) : S \subseteq T \subseteq V \text{ and } |T| = |S| + 1\}$$

with $\partial_-((S, T)) = S$ and $\partial_+((S, T)) = T$. Lastly, the weight of $S \in V'$ is $|\Theta(S)|$.

Lemma 2.1 *There is a one-to-one correspondence between s-t paths in $N(G)$ and numberings of G.*

Proof Given a numbering $\eta : V \to \{1, 2, ..., n\}$, the sets $S_0(\eta)$, $S_1(\eta)$, ..., $S_n(\eta)$ determine an s-t path in G'. Conversely, given an s-t path, S_0, S_1, ..., S_n, in G', the function $\eta : V \to \{1, 2, ..., n\}$ defined by

$$\eta^{-1}(k) = S_k - S_{k-1}, \; 1 \leq k \leq n,$$

is a numbering of G. □

Theorem 2.1 *The wirelength problem on G and the minpath problem on $N(G)$ are equivalent.*

Proof By the previous lemma, the weight of an s-t path, $S_0, S_1, ..., S_n$, in $N(G)$ is

$$\sum_{k=0}^{n} |\Theta(S_k)| = \sum_{k=0}^{n} |\Theta(S_k(\eta))| = wl(\eta)$$

by the identity of Section 1.3.1.2 of Chapter 1. □

We call $N(G)$ the *derived network of* G. What is the significance of this reduction? The wirelength problem is known to be intractable in general (NP-complete). The minpath problem does have a nice $O(m)$ algorithm for solving it when the graph is acyclic, which is the case for $N(G)$. However, $N(G)$ itself is exponentially larger than G. There is some savings in transforming the problem since the brute force solution of the wirelength problem on G requires evaluation of all $n!$ numberings and by Stirling's formula $n! \simeq \sqrt{2\pi n}\left(\frac{n}{e}\right)^n$. The complexity of the acyclic algorithm is only $O(|E'|)$ and $|E'| = n2^{n-1}$ (see Exercise 1.1 of Chapter 1). So the minpath problem on $N(G)$ has a more efficient solution for large n but it is still exponential in n and therefore impractical. The real significance of the reduction is the fact which we examine in the next section, that the minpath problem has a natural notion of morphism.

2.4 Pathmorphisms

2.4.1 Definitions

Before we proceed to the main definition of this chapter, we must define a preliminary notion, that of homomorphism for directed graphs. A *digraph morphism* $\varphi = (\varphi_V, \varphi_E) : G \to H$ consists of a pair of functions, $\varphi_V : V_G \to V_H$ and $\varphi_E : E_G \to E_H$, such that for all $e \in E_G$, $\partial_\pm(\varphi_E(e)) = \varphi_V(\partial_\pm(e))$, which is to say that they map vertices to vertices and edges to edges so as to preserve their incidences. Equivalently, in the language of category theory, it is that the diagram in Fig. 2.3 commutes.

A *pathmorphism*, $\varphi : M \to N$, from a network M to a network N (both vertex weighted), consists of a digraph morphism from M to N such that

(1) $\varphi(s_M) = s_N$ and $\varphi(t_M) = t_N$,
(2) for all $v \in V_M$, $\omega_M(v) \geq \omega_N(\varphi(v))$,

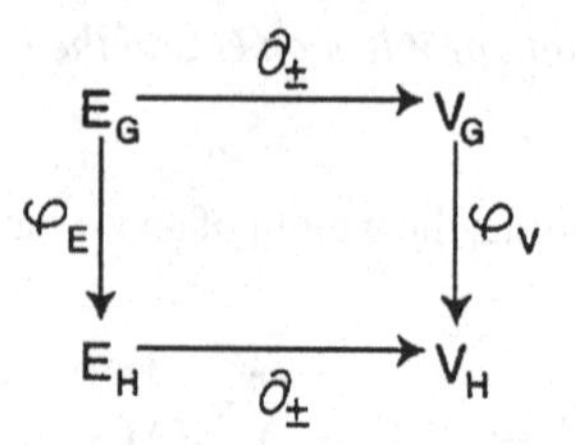

Fig. 2.3 A digraph morphism, $\varphi : G \to H$.

(3) φ has a right inverse, $\rho : N \to M$, having properties 1 and 2 (i.e. $\varphi \circ \rho = \iota_N$, the identity on N).

Theorem 2.2 *(The Fundamental lemma) If $\varphi : M \to N$ is a pathmorphism, then*

$$\min_{P \in \mathfrak{P}(M)} \omega(P) = \min_{P \in \mathfrak{P}(N)} \omega(P).$$

Proof From the definition of a digraph morphism and (1), any s-t path in M, with vertices $s_M, v_1, \ldots, v_k, t_M$, will be mapped to an s-t path in N, with vertices $s_N = \varphi(s_M), \varphi(v_1), \ldots, \varphi(v_k), \varphi(t_M) = t_N$. By (2), the total weight of the image path will be no more than that of the original path in M. N may also contain s-t paths other than those which are images of paths in M. In any case,

$$\min_{P \in \mathfrak{P}(M)} \omega(P) \geq \min_{P \in \mathfrak{P}(N)} \omega(P)$$

and by (3) we have the opposite inequality which proves the theorem. □

Thus, the minimum path problems on M and N are equivalent (even though N is essentially a subnetwork of M, and may be much smaller). This theorem is called the Fundamental lemma since it shows that pathmorphisms preserve the minpath problem. We call any theorem which states that a certain set of transformations preserves a certain problem *the Fundamental lemma* (for those transformations and that problem).

2.4.2 Examples

Fig. 2.4 shows a digraph homomorphism which is a pathmorphism and Fig. 2.5 shows a digraph homomorphism which is not a pathomorphism.

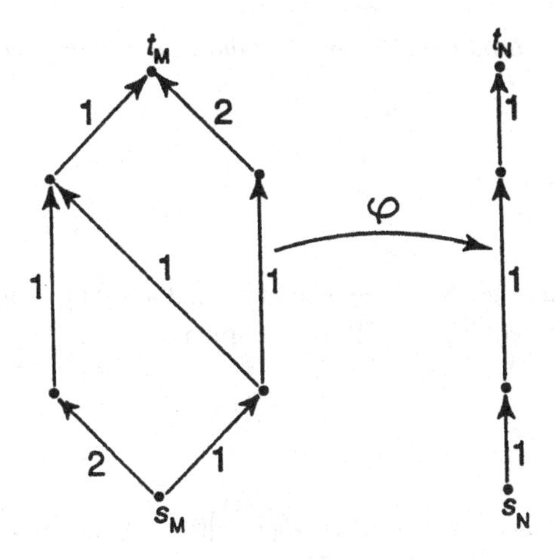

Fig. 2.4

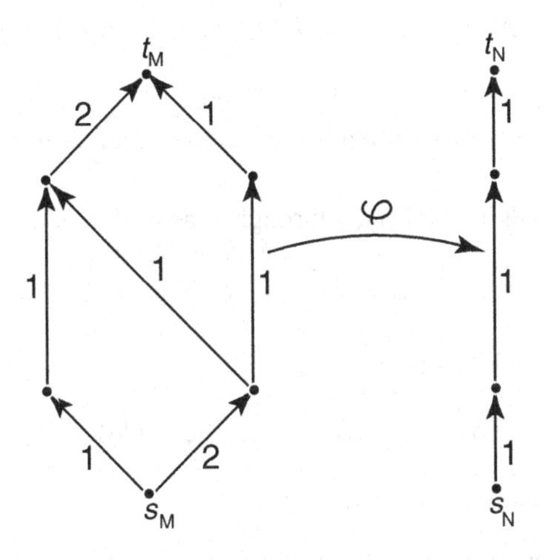

Fig. 2.5

Exercise 2.3 *Explain how the mapping, φ, of Fig. 2.5 does not satisfy the definition of a pathmorphism (which it does not, since* $\min_{P\in\mathfrak{P}(M)}\omega(P)=4$ *and* $\min_{P\in\mathfrak{P}(N)}\omega(P)=3$*).*

2.4.3 Steiner operations

A *Steiner operation* on a graph, G, is a pathmorphism on the derived network, $N(G)$. Since this is a rather roundabout definition, let us characterize Steiner operations more directly.

Theorem 2.3 *A set-map $\varphi : 2^V \to 2^V$ induces a Steiner operation iff $\forall S \subseteq T \subseteq V$,*

(1) $|\varphi(S)| = |S|$,
(2) $|\Theta(\varphi(S))| \leq |\Theta(S)|$ *and*
(3) $\varphi(S) \subseteq \varphi(T)$.

Proof Suppose that $\varphi : N(G) \to N(G)$ is a pathmorphism and that $\varphi_{V'}$ is its vertex-component. $\varphi_{V'} : 2^V \to 2^V$ is a set-map on V and $S \subseteq T \subseteq V$. Then

(1) Let

$$\emptyset = S_0 \subset S_1 \subset \ldots \subset S_k = S$$

be an $\emptyset$-S path in $N(G)$. Then $|S_0| = |\emptyset| = 0$ and by induction $|S_k| = |S_{k-1}| + 1 = (k-1) + 1 = k$. The image of that path under $\varphi_{V'}$ is

$$\emptyset = \varphi_{V'}(S_0) \subset \varphi_{V'}(S_1) \subset \ldots \subset \varphi_{V'}(S_k) = \varphi_{V'}(S)$$

with $\left|\varphi_{V'}(S_k)\right| = \left|\varphi_{V'}(S_{k-1})\right| + 1$. Therefore $\left|\varphi_{V'}(S)\right| = k$ also.

(2) $\left|\Theta\left(\varphi_{V'}(S)\right)\right| \leq |\Theta(S)|$ is equivalent to Condition 2 in the definition of a pathmorphism.

(3) Letting the path in $N(G)$ pass through T as well as S we have

$$S = S_k \subset \ldots \subset S_l = T,$$

whose image under $\varphi_{V'}$ is

$$\varphi_{V'}(S) = \varphi_{V'}(S_k) \subset \ldots \subset \varphi_{V'}(S_l) = \varphi_{V'}(T)$$

and Part 3 follows by transitivity of $\subset$.

Conversely, suppose that we have a set-map, φ, satisfying (1), (2) and (3). Let $\varphi_{V'} = \varphi$. Then by (1), $\varphi_{V'}(s_M) = \varphi_{V'}(\emptyset) = \emptyset = s_N$ and similarly $\varphi_{V'}(t_M) = t_N$. Also, define $\varphi_{E''}$ on $N(G)$ by

$$\varphi_{E''}(S, T) = \left(\varphi_{V'}(S), \varphi_{V'}(T)\right).$$

By (1) and (3) $\varphi_{E''}(S, T) \in E'$ so φ has induced a digraph homomorphism on $N(G)$. The range of this digraph homomorphism will be a subnetwork M of $N(G)$ and so ρ can be the embedding map. Again, the second part of both definitions are essentially the same. □

Theorem 2.3 does make Steiner operations more concrete but we are still lacking examples. We address that issue in the next chapter.

2.5 Comments

Writing a thorough history of minpath problems is beyond my limited scholarship but I will take this opportunity to put forth some observations. The origins of the subject are probably as old as humanity itself. Trailblazer, pathfinder, shortcut are part of the language. The human brain evolved to be able to "make connections." In a number of ancient cultures, e.g. in Minoan, the maze was a central metaphor. The well-known puzzle of the Man, Wolf, Goat and Cabbage, essentially a minpath problem, is probably prehistoric.

Much of classical mathematics implicitly involves minpath problems. In algebra, a solution of a polynomial equation

$$\sum_{k=0}^{n} a_k x^k = 0,$$

is a series of transformations of the equation by arithmetic operations and the taking of radicals until it is in the form

$$x = F(a_0, a_1, ..., a_n).$$

Similarly for the solution of differential equations and the reduction of matrices to canonical form. A mathematical proof is, ideally, a sequence of valid implications from axioms or theorems to the statement in question.

The calculus of variations, from Newton to Euler, was about a continuous analog of the acyclic minpath problem (see [**40**]). This may not be apparent since most books on the subject (e.g. [**38**]) emphasize the variational method which is analogous to differentiation. The functions being minimized over are therefore thought of as "points" in an infinite-dimensional space. However, the graph of such a function, $y = f(x)$, defined on an interval, $[a, b]$, is a curve from $s = (a, f(a))$ to $t = (b, f(b))$ and the function to be minimized

$$\omega(f) = \int_a^b F\left(x, y, y', ..., y^{(n)}\right) \mathrm{d}x,$$

is the analog of $\omega(P)$, the weight of a path. The brachystochrone problem, posed, but incorrectly solved, by Galileo, is the classic example.

Steiner operations were named after J. Steiner, the great Swiss mathematician of the mid-nineteenth century. Our intent is to bring out the analogy with symmetrization, an operation which Steiner defined on sets in the plane and used to prove the classical isoperimetric theorem of Greek geometry (see the appendix). Symmetrization was subsequently applied by Lord Rayleigh and others to problems of mathematical physics. A typical application was to show that the drumhead with a fixed area having the lowest fundamental frequency is circular. A good reference for applications and variations of symmetrization is the book *Isoperimetric Inequalities in Mathematical Physics* by Polya and Szego [**84**]. Steiner's insight into the classical isoperimetric problem was that symmetrization preserves area and does not increase the length of the boundary of a set in the plane. Properties (1) and (2) of Theorem 2.3 are the analogs of those key properties. Though it is never explicitly mentioned, Steiner symmetrization and all of its variants in Polya and Szego's book also have property (3). In a continuum, property (3) is tantamount to continuity.

It has been of considerable solace to the author that several of the most famous blunders in mathematical history occurred in the calculus of variations, demonstrating the subtlety and difficulty of this subject so closely related to that of our monograph. The aforementioned mistake by Galileo on the brachystochrone shows that sometimes just guessing the solution of a variational problem is not easy. Of course those are the most interesting ones, but even when a solution is easy to guess it may still be difficult to prove. The classical isoperimetric theorem did not have a proof for several thousand years. Steiner published a proof in 1840 and refused to accept the fact that it contained a logical gap for some weeks after being pointed out by Weierstrass. A discussion of this incident is to be found on p. 58 of [**60**]. The fact is, isoperimetric theorems, whether combinatorial or continuous, are difficult to prove, even when easy to guess.

In the modern era, Richard Bellman was a leader in applying the computer to minimum path problems. Unfortunately, in the opinion of this author, he made some poor choices. The main one was in not completely abstracting his concepts from the applications which suggested them. He wrote about (see [**9**]) "multistage decision processes" with "states," "decisions" and "policies" and called the subject "dynamic programming." His multistage decision processes are actually networks: Their states represent vertices, decisions represent edges and policies are then s-t paths. So "Dynamic programming" is just another name for the minpath problem. The effect of such nonstandard nomenclature is to obscure the subject and cut it off from the rest of mathematics. To illustrate the consequences, try doing arithmetic in a foreign language, even one which you speak well. Bellman's nomenclature has been largely abandoned within combinatorial mathematics but still persists in applied areas and OR (operations research). Bellman correctly identified the "optimality principle" (the idea behind Bellman's equations) as fundamental and found the acyclic algorithm, but those were surely known to previous generations of mathematicians, folk theorems rather than new discoveries. The vertex set in most of Bellman's examples is infinite, there being a continuum of them at each stage. Evidently, he was unable to completely let go of his training as an analyst and embrace the combinatorial. On the other hand, he may have also been ahead of his time. It seems likely that, in the not-too-distant future, there will be a need for a numerical analysis of minpath problems and then Bellman's notion of "convergence in policy space" could yet prove valuable.

3
Stabilization and compression

3.1 Introduction

In the literature on combinatorial isoperimetric problems on a graph G, there are two systematic ways in which Steiner operations have been constructed:

(1) *stabilization*, based on certain kinds of reflective symmetry of G, and
(2) *compression*, based on product decompositions of G with certain kinds of factors.

3.2 Stabilization

3.2.1 Diagrams

Recall that a diagram is a graph whose vertices are points in $\mathbb{R}^d$, d-dimensional Euclidean space, and whose edges are arcs connecting pairs of vertices. The edges need not be straight lines, though they may always be taken as such, and may even intersect at interior points. In Chapter 1 planar diagrams were just used to give a visual representation of graphs but now we wish to take advantage of the geometry of the ambient space, so let us define these representations more carefully.

3.2.1.1 *The n-gon,* $\mathbb{Z}_n$

The vertices of $\mathbb{Z}_n$ are the nth roots of unity (in $\mathbb{R}^2$),

$$v_k = (\cos 2\pi k/n, \sin 2\pi k/n), \quad k = 0, 1, ..., n-1,$$

and its edges are circular arcs

$$e_k = \{(\cos 2\pi (k+t)/n, \sin 2\pi (k+t)/n) : 0 \leq t \leq 1\},$$
$$k = 0, 1, ..., n-1,$$

with $\partial (e_k) = \left\{v_k, v_{k+1(\mathrm{mod}\, n)}\right\}.$

3.2.1.2 *The d-cube,* $\mathbb{Q}_d$

The vertices are $V_{Q_d} = \{-1, 1\}^d \subset \mathbb{R}$. The edges,

$$E_{Q_d} = \left\{(x, y) \in \binom{V}{2} : \exists i_0 \text{ such that } x_i = y_i \text{ for } i \neq i_0 \text{ and } x_{i_0} \neq y_{i_0}\right\}$$

are represented by the straight-line segments determined by each such pair, i.e. for $e \in V_{Q_d}$ with $\partial(e) = \{x, y\}$, e is represented by the arc

$$\{tx + (1 - t)y : 0 \leq t \leq 1\}.$$

3.2.1.3 *The d-crosspolytope,* $\maltese_d$

Let $\delta^{(k)}$ be the d-tuple with all entries 0 except the kth, which is 1. The vertex set of $\maltese_d$ will then be $V_{\maltese_d} = \left\{\pm\delta^{(k)} : k = 1, 2, ..., d\right\}$. All pairs of these points, except the antipodal pairs $\left\{\delta^{(k)}, -\delta^{(k)}\right\}$, $k = 1, ..., d$, are connected with straight-line edges (see [**28**], pp. 121–2). $\maltese_2$ is the square again and $\maltese_3$ is commonly known as the octahedron.

3.2.2 Symmetries

A distance-preserving linear transformation of $\mathbb{R}^d$ onto itself is called an *orthogonal transformation.* The group of all orthogonal transformations is called the *orthogonal group* and denoted by O_d. A *reflection,* $\mathcal{R}$, on $\mathbb{R}^d$ is an orthogonal transformation which keeps a hyperplane (the solution set of a linear equation, $e \cdot x = 0$) fixed and maps every other point to its mirror image in the hyperplane. If e is a unit vector, then the image of x is

$$\mathcal{R}(x) = x - 2(e \cdot x)e.$$

Note that a hyperplane cuts the space into two components, and that every arc from one component to the other must intersect the hyperplane.

A *linear automorphism* of a diagram is an orthogonal transformation of $\mathbb{R}^d$ which acts as an automorphism of the graph, i.e. it maps vertices to vertices and edges to edges. If, as we may assume, the vertices of G span the whole space, the linear automorphisms of a diagram form a group, a subgroup of O_d, and it is finite.

3.2.3 Examples

3.2.3.1 *The dihedral group,* D_n

D_n consists of rotations by $2\pi k/n$ radians, $0 \leq k < n$, about the origin, and reflections about the lines $\theta = \pi k/n$, $0 \leq k < n$. $|D_n| = 2n$ but it may be generated by just two reflections, such as those about the lines $\theta = 0$ and $\theta = \pi/n$ (note that the composition of two reflections is a rotation). D_n is the symmetry group of $\mathbb{Z}_n$.

3.2.3.2 *The cuboctahedral group*

Any linear transformation is determined by its action on a basis. The positive vertices $\{\delta^{(k)} : k = 1, 2, ..., d\}$ of $\maltese_d$ constitute a basis of $\mathbb{R}^d$. So if F is a linear automorphism of $\maltese_d$, $F(\delta^{(1)})$ may be any of the $2d$ vertices of $\maltese_d$, say $\pm\delta^{(k)}$ and then by linearity $F(-\delta^{(1)}) = -F(\delta^{(1)}) = \mp\delta^{(k)}$. The remaining $d - 1$ basis members may then be recursively mapped onto the remaining $2(d-1)$ vertices in the same manner and these choices uniquely determine F. The group of linear automorphisms of $\maltese_d$ is thus of order $2^d d!$ and it is transitive on vertices and edges (see [**28**], p. 133). Also, the transpositions $\delta^{(k)} \longleftrightarrow -\delta^{(k)}$, $k = 1, 2, ..., d$, and $\delta^{(j)} \longleftrightarrow \pm\delta^{(k)}$, $1 \le j < k \le d$, correspond to reflections which generate the group. This same group constitutes the linear automorphisms of the d-cube, where it is also transitive on vertices and edges.

3.2.4 Definition

Given a diagram G in $\mathbb{R}^d$, a *stabilizing reflection* $\mathcal{R}$, of G, is a reflection (of $\mathbb{R}^d$) which

(1) acts as a linear automorphism of G, taking vertices to vertices and edges to edges, and
(2) if $e \in E$, $\partial(e) = \{v, w\}$ with v and w on different sides (components) of the fixed hyperplane of $\mathcal{R}$, then $\mathcal{R}(e) = e$, i.e. $\mathcal{R}(v) = w$ and $\mathcal{R}(w) = v$.

Theorem 3.1 *If G is a diagram in $\mathbb{R}^d$ and $\mathcal{R}$ is a reflective symmetry of G which is not stabilizing, then there are distinct edges (arcs in $\mathbb{R}^d$) which have a common interior point.*

Proof By the definition of a stabilizing reflection, there exists an edge, e, distinct from $\mathcal{R}(e)$. If t is a point on e then $\mathcal{R}(t)$ is on $\mathcal{R}(e)$. Since e is an arc connecting v to w which are on opposite sides of the fixed hyperplane of $\mathcal{R}$, there is a point, t_0, in the interior of e which lies on the hyperplane. Therefore $t_0 = \mathcal{R}(t_0) \in \mathcal{R}(e)$ and so $t_0 \in e \cap \mathcal{R}(e)$. □

Since edges of $\mathbb{Z}_n$ intersect only at vertices, all reflections in the dihedral group are stabilizing. Similarly, all reflections in the cuboctahedral group are stabilizing for both $\maltese_d$ and Q_d.

With a stabilizing symmetry, $\mathcal{R}$, of G, and a point $p \in \mathbb{R}^d$ (called the Fricke–Klein point) such that $\mathcal{R}(p) \neq p$ (i.e. p is not on the fixed hyperplane of $\mathcal{R}$), we may define the operation of *stabilization* on subsets of V. For $S \subseteq V$, let

$$S' = \{v \in S : \mathcal{R}(v) \notin S \text{ and } \|v - p\| > \|\mathcal{R}(v) - p\|\}.$$

Here "$\|\cdot\|$" denotes the usual Euclidean metric. By the triangle inequality, "$\|v - p\| > \|\mathcal{R}(v) - p\|$" means that $\mathcal{R}(v)$ is on the same side of the fixed hyperplane as p and v is on the opposite side. Then let

$$S'' = S - S'$$

or equivalently

$$S = S' + S'' \text{ (disjoint union)},$$

and define

$$Stab_{\mathcal{R},p}(S) = \mathcal{R}(S') + S''.$$

In other words, $Stab_{\mathcal{R},p}(S)$, *the stabilization of S* with respect to (wrt) *R and* p, consists of the symmetrical portion of S together with all p-side vertices v such that either v or $\mathcal{R}(v)$ belongs to S.

Example 3.1 *See Figs. 3.1 and 3.2. The dotted line ($\theta = 0$) in these figures is the fixed line of a reflection, $\mathcal{R}$. Members of S are darkened.*

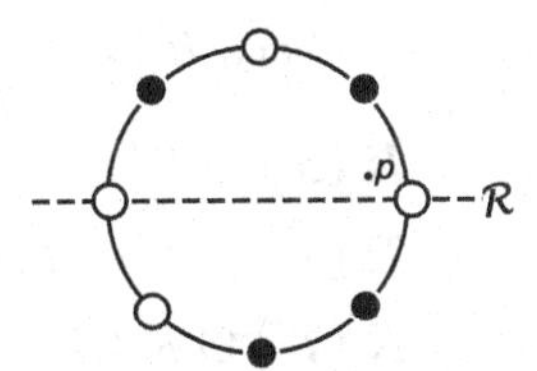

Fig. 3.1 A set $S \subset \mathbb{Z}_8$.

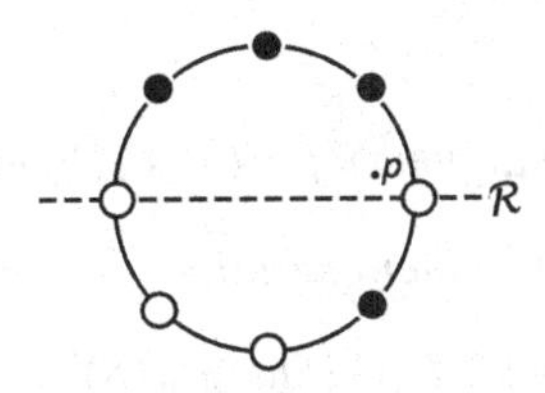

Fig. 3.2 $Stab_{\mathcal{R},p}(S)$.

3.2.5 Basic properties of stabilization

Theorem 3.2 *For all $S, T \subseteq V$,*

(1) $\left|Stab_{\mathcal{R},p}(S)\right| = |S|$,
(2) $\left|\Theta\left(Stab_{\mathcal{R},p}(S)\right)\right| \leq |\Theta(S)|$,
(3) *$S \subseteq T$ implies $Stab_{\mathcal{R},p}(S) \subseteq Stab_{\mathcal{R},p}(T)$.*

Proof

(1)
$$\begin{aligned}\left|Stab_{\mathcal{R},p}(S)\right| &= \left|\mathcal{R}(S') + S''\right| \\ &= \left|\mathcal{R}(S')\right| + \left|S''\right| \\ &= \left|S'\right| + \left|S''\right| \\ &= \left|S' + S''\right| \\ &= |S|.\end{aligned}$$

(2) Now consider the edges in $\Theta\left(Stab_{\mathcal{R},p}(S)\right)$ but not in $\Theta(S)$. For each such edge we shall find a unique edge which is in $\Theta(S)$ but not in $\Theta\left(Stab_{\mathcal{R},p}(S)\right)$. If $e \notin \Theta(S)$ and $\partial(e) = \{v, w\}$, then either
   (a) $\{v, w\} \subseteq S$: By the definition of stabilizing reflection, v and w must be on the same side of the fixed hyperplane of $\mathcal{R}$. One of them, say v, must be in S'. This means that v, and therefore w, must be on the non-p side of the fixed hyperplane. Also, w must be in S''. But then $\mathcal{R}(v) \notin S$ and $\mathcal{R}(w) \in S$ which means that $\mathcal{R}(e) \in \Theta(S)$ but $\mathcal{R}(v) \in Stab_{\mathcal{R},p}(S)$ so $\mathcal{R}(e) \notin \Theta\left(Stab_{\mathcal{R},p}(S)\right)$.
   (b) $\{v, w\} \cap S = \emptyset$.

(3) Note that $S \subseteq T$ implies $S'' \subseteq T''$ and $\mathcal{R}\left(S' - T'\right) \subseteq T''$. Then

$$\begin{aligned}Stab_{\mathcal{R},p}(S) &= \mathcal{R}\left(S'\right) + S'' \\ &\subseteq \mathcal{R}\left(T' \cup \left(S' - T'\right)\right) \cup T'' \\ &= \mathcal{R}\left(T'\right) \cup \mathcal{R}\left(S' - T'\right) \cup T'' \\ &= \mathcal{R}\left(T'\right) + T'' = Stab_{\mathcal{R},p}(T).\end{aligned}$$

□

Exercise 3.1 *Write out the argument for Case 2(b) of Theorem 3.2.*

Corollary 3.1 *$Stab_{\mathcal{R},p}$ is a Steiner operation.*

Note that in Figs 3.1 and 3.2 $\left|\Theta\left(Stab_{\mathcal{R},p}(S)\right)\right| = 4 < 6 = |\Theta(S)|$, so the inequality of Theorem 3.2(2) may be strict.

3.2.6 Multiple stabilizations

If T is in the range of $Stab_{\mathcal{R},p}$, i.e. $T = Stab_{\mathcal{R},p}(S)$ for some S, then since $Stab_{\mathcal{R},p}(S)' = \emptyset$,

$$Stab_{\mathcal{R},p}(T) = Stab_{\mathcal{R},p}\left(Stab_{\mathcal{R},p}(S)\right) = Stab_{\mathcal{R},p}(S) = T.$$

A set, T, such that $Stab_{\mathcal{R},p}(T) = T$ is called *stable with respect to $\mathcal{R}$ and p*. The range of $Stab_{\mathcal{R},p}$ is thus induced by its stable sets. We have shown that in solving the wirelength problem on a diagram, G, having a stabilizing reflection, $\mathcal{R}$, we need not consider all of $N(G)$ but only the range of $Stab_{\mathcal{R},p}$.

Now suppose that G has $k > 1$ stabilizing reflections, $\mathcal{R}_0, \mathcal{R}_1, ..., \mathcal{R}_{k-1}$. Each simplifies our problem by mapping $N(G)$ to a subnetwork, M_i, but can we combine these individual Steiner operations into one, call it *Stab*, which embodies all their simplifications, i.e. such that $Stab_{\mathcal{R}_i,p_i} \circ Stab = Stab$ for $i = 0, 1, ..., k-1$? In category theory, the general theory of morphisms, if such a thing exists and has an additional technical property called "universality", it is called a "pushout" (see [**76**]). In general, pathmorphisms do not have pushouts, but for stabilization operations, providing we choose a common Fricke–Klein point, we can construct what is essentially a pushout. It does, however, lack universality!

Choose a point, p, not lying on the fixed hyperplanes of $\mathcal{R}_0, \mathcal{R}_1, ..., \mathcal{R}_{k-1}$ and consider the corresponding stabilizations, $Stab_{\mathcal{R}_i,p} : N(G) \to M_i \subseteq N(G)$. Given any set $S \subseteq V$, define a sequence of sets, $T_0, T_1, ..., T_j, ...$ by $T_0 = S$ and $T_{j+1} = Stab_{\mathcal{R}_{j(\mathrm{mod}\,k)},p}\left(T_j\right)$. To illustrate this, we take the set of four vertices on $\mathbb{Z}_8$ (the darkened ones) in Fig. 3.1. For reflections we select the generating reflections $\mathcal{R}_1$, having fixed line $\theta = 0$, and $\mathcal{R}_2$, having fixed line $\theta = \frac{\pi}{8}$. As we alternately apply stabilizations to successive sets, we follow their evolution in Fig. 3.3. p is in the first quadrant and between the two fixed lines.

Note that for $j \geq 5$, T_j is stable with respect to $\mathcal{R}_1, \mathcal{R}_2$ and p. This is always the case for j sufficiently large as the following lemma shows.

Lemma 3.1 *For all $S \subseteq V$, the sequence $T_0, T_1, ...$ is eventually constant, i.e. stable with respect to $\mathcal{R}_1, \mathcal{R}_2, ..., \mathcal{R}_k$ and p.*

Proof Let $\kappa(S) = \sum_{v \in S} \|v - p\|$. Then for any $\mathcal{R}_i$, $\kappa\left(Stab_{\mathcal{R}_i,p}(S)\right) \leq \kappa(S)$, with equality if and only if S is stable with respect to $\mathcal{R}_i$ and p. Also, if for any i, $T_i = T_{i+1} = ... = T_{i+k}$, then T_i is stable with respect to $\mathcal{R}_1, \mathcal{R}_2, ..., \mathcal{R}_k$ and p. Each time $\kappa\left(T_j\right)$ does decrease, it must be by at

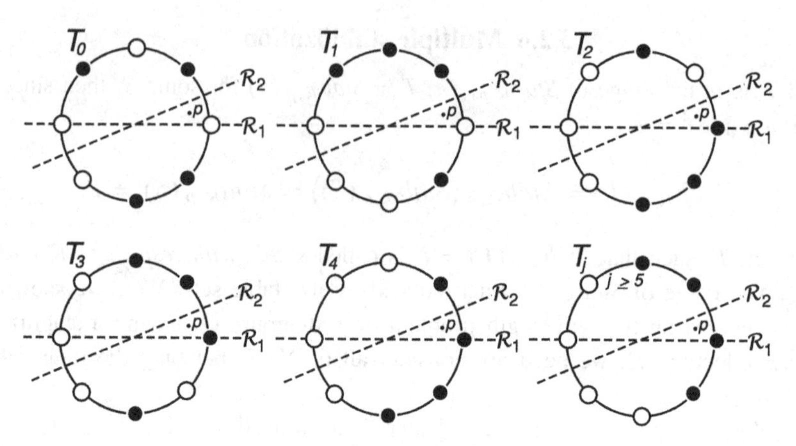

Fig. 3.3 Six stages of stabilization for S.

least $\epsilon = \min \{\|v - p\| - \|\mathcal{R}(v) - p\| > 0 : v \in V\}$. Therefore $\kappa(T_j)$ must be constant for j sufficiently large ($j > (k-1)\kappa(S)/\epsilon$ will do it). □

$(n-1)\kappa(V)/\epsilon$ is a bound which works for all $S \subseteq V$ so we may define a Steiner operation

$$Stab^{(\infty)} = Stab_{\mathcal{R}_{N(\mathrm{mod}\,k)},p} \circ Stab_{\mathcal{R}_{N-1(\mathrm{mod}\,k)},p} \circ \ldots \circ Stab_{\mathcal{R}_1,p}.$$

This is not quite a pushout for the $Stab_{\mathcal{R}_i,p}$, $0 \leq i < k$, since it may depend on the order of the $\mathcal{R}_i$ (and thus is not universal), but it does have the property that $Stab_{\mathcal{R}_i,p} \circ Stab^{(\infty)} = Stab^{(\infty)}$, $0 \leq i < k$. All we need to know then is the range of $Stab^{(\infty)}$, i.e. the subnetwork of $N(G)$ induced by sets stable with respect to all of $\mathcal{R}_1, \mathcal{R}_2, \ldots, \mathcal{R}_k$ and p. But how can we determine which subsets of V are stable without checking through all of them (an exponential process which would defeat our whole purpose)?

3.2.7 Stability order

A *partially ordered set* (poset), $\mathcal{P} = (P, \leq)$, consists of a set, P, and a binary relation, $\leq$, on P, which is

(1) reflexive: $\forall x \in P, x \leq x$,
(2) antisymmetric: $\forall x, y \in P$, if $x \leq y$ and $y \leq x$ then $x = y$, and
(3) transitive: $\forall x, y, z \in P$, if $x \leq y$ and $y \leq z$ then $x \leq z$.

3.2.7.1 Definition

Let

$$P^{(0)} = \{(v, v) : v \in V\}, \text{ the identity on } V,$$

$$P^{(1)} = \bigcup_{i=0}^{k-1} \{(v, w) : \mathcal{R}_i(v) = w \text{ and } \|v - p\| < \|w - p\|\}, \text{ and}$$

$$P^{(j)} = P^{(1)} \circ P^{(j-1)} \text{ for } j > 1.$$

Then we define

$$\mathcal{S}(G; \mathcal{R}_1, \mathcal{R}_2, ..., \mathcal{R}_k; p) = \bigcup_{j=0}^{\infty} P^{(j)}.$$

By definition $\mathcal{S}(G; \mathcal{R}_1, \mathcal{R}_2, ..., \mathcal{R}_k; p)$ is reflexive and transitive; in fact it is the reflexive and transitive closure of $P^{(1)}$. That it is antisymmetric, and thus a partial order, follows from the observation that $(v, w) \in P^{(1)}$ means that v is strictly closer to p than w is. $\mathcal{S}$ will be called the *stability order of G with respect to $\mathcal{R}_1, \mathcal{R}_2, ..., \mathcal{R}_k$ and p* and if $(v, w) \in \mathcal{S}$ we shall write $v \leq_{\mathcal{S}} w$. $P^{(1)} - \bigcup_{j>1} P^{(j)}$ is the smallest subset of $P^{(1)}$ whose reflexive and transitive closure is still $\mathcal{S}$. The digraph with vertices V and edges $P^{(1)} - \bigcup_{j>1} P^{(j)}$ is known as the *Hasse diagram* of $\mathcal{S}$.

3.2.7.2 Examples

(1) The n-gon, $\mathbb{Z}_n$: Let $k = 2$ and $\mathcal{R}_1, \mathcal{R}_2$ be the generators of D_n mentioned in Example 3.2.3.1, i.e. $\mathcal{R}_1$ is reflection about the x-axis ($\theta = 0$) and $\mathcal{R}_2$ about the line $\theta = \pi/n$. Choose p in the sector $0 < \theta < \pi/n$. The stability order $\mathcal{S}(\mathbb{Z}_n; \mathcal{R}_1, \mathcal{R}_2; p)$ is then a total order

$$v_0 <_{\mathcal{S}} v_1 <_{\mathcal{S}} v_{-1} <_{\mathcal{S}} v_2 <_{\mathcal{S}} v_{-2} <_{\mathcal{S}} ...,$$

where $v_k = (\cos 2\pi k/n, \sin 2\pi k/n)$. To verify this, observe that $\mathcal{R}_2(v_1) = v_0$ and $\|v_1 - p\| > \|v_0 - p\|$; $\mathcal{R}_1(v_{-1}) = v_1$ and $\|v_{-1} - p\| > \|v_1 - p\|$; and so on.

(2) The crosspolytope, $\maltese_d$: Choose $p = \left(2^{-1}, 2^{-2}, ..., 2^{-d}\right) \in \mathbb{R}$. For any pair of vertices, $\pm\delta^{(j)}, \pm\delta^{(k)}$ there is a unique reflection which interchanges them, and all reflections are stabilizing. The stability order is therefore total in this case too, the order being given by distance from p:

$$\delta^{(1)} <_{\mathcal{S}} \delta^{(2)} <_{\mathcal{S}} ... <_{\mathcal{S}} \delta^{(d)} <_{\mathcal{S}} -\delta^{(d)} <_{\mathcal{S}} ... <_{\mathcal{S}} -\delta^{(1)}$$

since

$$\left\|\delta^{(k)} - p\right\| = \left(-\frac{1}{2}\right)^2 + \left(-\frac{1}{4}\right)^2 + \ldots + \left(1 - \frac{1}{2^k}\right)^2 + \ldots + \left(-\frac{1}{2^d}\right)^2$$
$$= \frac{1}{4}\frac{1 - \frac{1}{2^{2d}}}{1 - \frac{1}{4}} + 1 - \frac{2}{2^k}$$
$$= \frac{1}{3}\left(1 - \frac{1}{2^{2d}}\right) + 1 - \frac{2}{2^k},$$

and similarly

$$\left\|-\delta^{(k)} - p\right\| = \frac{1}{3}\left(1 - \frac{1}{2^{2d}}\right) + 1 + \frac{2}{2^k}.$$

(3) The d-cube, Q_d: The symmetries and p are the same as in the previous example (the cuboctahedral group) but the stability order is qualitatively different. If $\mathcal{R}_i$ is the reflection which negates the ith coordinate and $v \in V$ with $v_i = 1$, then $v <_{\mathcal{S}} \mathcal{R}_i(v)$ since $\|v - p\| < \|\mathcal{R}_i(v) - p\|$. Changing our representation of V_{Q_d} will help bring out the pattern here: The components of the d-tuple, x, of 0s and 1s will be the exponents, $v_i = (-1)^{x_i}$, of the components of v. Then $x <_{\mathcal{S}} \mathcal{R}_i(x) = y$ means that $y_k = x_k$ for all $k \neq i$, $x_i = 0$ and $y_i = 1$. Thus the Boolean lattice, $\mathcal{B}_d = \{0 < 1\}^d$, is a suborder of $\mathcal{S}$. The reflection, $\mathcal{R}_{ij}$, which interchanges the ith and jth components ($i < j$), will send x to y where $y_k = x_k$ for all $k \neq i, j$, $y_i = x_j$ and $y_j = x_i$. If $x_i = 0$ and $x_j = 1$, then $x <_{\mathcal{S}} y = \mathcal{R}_{i_j}(x)$. Thus we go up in $\mathcal{S}$ by "shifting 1s to the left and 0s to the right". For $d = 2$, $\mathcal{S}$ is a total order,

$$(0,0) <_{\mathcal{S}} (0,1) <_{\mathcal{S}} (1,0) <_{\mathcal{S}} (1,1),$$

but for $d = 3$ it is not. Its Hasse diagram is shown in Fig. 3.4.

For $d = 4$ its Hasse diagram is shown in Fig. 3.5.

In addition to choosing a Fricke–Klein point not on the fixed hyperplane of a stabilizing symmetry, it is convenient to choose p so that the distances $\|v - p\|$, $v \in V$, are all distinct. This gives a total order on V (by increasing distance from p) which we call the *Fricke–Klein order*.

Exercise 3.2 *Show that the Fricke–Klein order on Q_d (with the given p) is lexicographic.*

3.2.8 Ideals

An *ideal* (also called a lower set or down set in the literature) in a poset, $\mathcal{P} = (P, \leq)$, is a set $S \subseteq P$ such that $x \leq y$ and $y \in S$ imply $x \in S$.

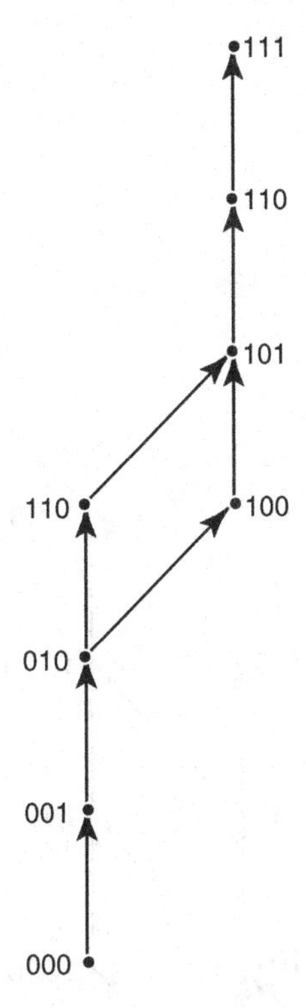

Fig. 3.4 The stability order of Q_3.

Theorem 3.3 *$S \subseteq V$ is a stable set wrt $\mathcal{R}_1, \mathcal{R}_2, ..., \mathcal{R}_k$ and p if and only if S is an ideal in the stability order, $\mathcal{S}(G; \mathcal{R}_1, \mathcal{R}_2, ..., \mathcal{R}_k; p)$.*

Proof Suppose that S is an ideal in $\mathcal{S}(G; \mathcal{R}_1, \mathcal{R}_2, ..., \mathcal{R}_k; p)$ but $\exists i$ such that S is not stable wrt $\mathcal{R}_i$. Then $\exists y \in S$ such that $\mathcal{R}_i(y) \notin S$ and $\|\mathcal{R}_i(y) - p\| < \|y - p\|$. But then $\mathcal{R}_i(y) <_S y$, which is a contradiction. Conversely, suppose that $x <_S y \in S$. By the definition of $\mathcal{S}(G; \mathcal{R}_1, \mathcal{R}_2, ..., \mathcal{R}_k; p)$, $\exists x_0, x_1, ..., x_n \in V$ such at $x = x_0$, $x_n = y$ and for $1 \leq i \leq n$ $\exists j_i$ such that $\mathcal{R}_{j_i}(x_{i-1}) = x_i$ and $\|x_{i-1} - p\| < \|x_i - p\|$. Then $x_n = y \in S$ implies that $x_{n-1} \in S$ which implies that $x_{n-2} \in S$ and so on until $x = x_0 \in S$. $\square$

This theorem reduces the problem of identifying stable sets to that of calculating the stability order and identifying ideals in it.

3.2.9 The derived network

Having calculated $\mathcal{S}(G; \mathcal{R}_1, \mathcal{R}_2, ..., \mathcal{R}_k; p)$ in the above examples, it is easy to represent the range of $Stab^{(\infty)}$.

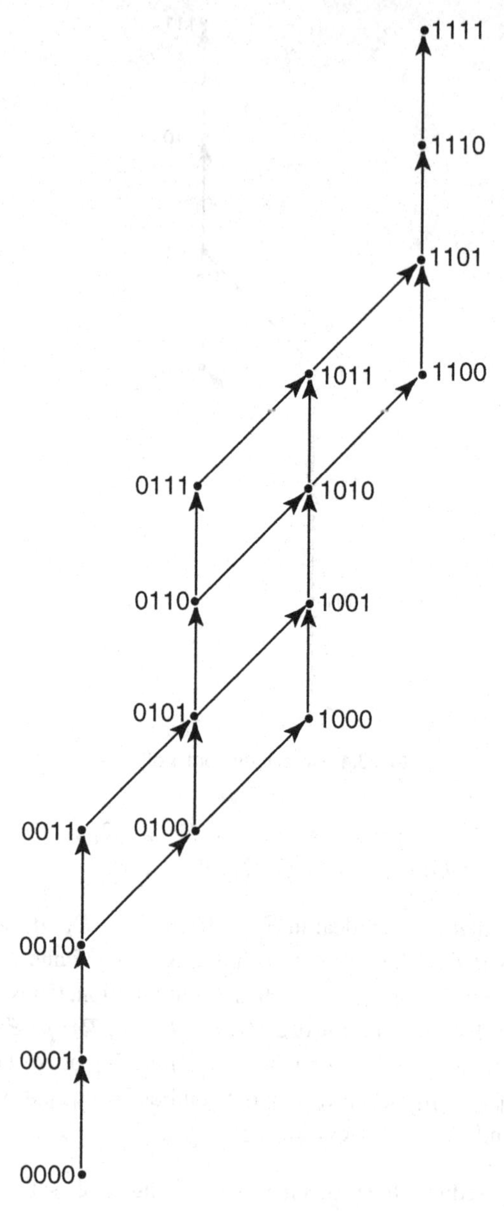

Fig. 3.5 Hasse diagram of $\mathcal{S}(Q_4)$.

3.2.9.1 Definition

The *derived network,* $N(\mathcal{S})$, is just the subnetwork of $N(G)$ induced by the ideals of $\mathcal{S}$. Thus $s = \emptyset$ and $t = V$ and an edge connects S to T if and only if $S \subset T$ and $|T| = |S| + 1$. The weight of an ideal S is $|\Theta(S)|$.

3.2.9.2 Examples

(1) The n-gon, $\mathbb{Z}_n$, the d-crosspolytope, $\maltese_d$ and square, Q_2: They all have stability orders which are total. Since the ideals of a total order are themselves totally ordered by containment, the derived network of a total order is just a single s-t path. The optimal numbering for the graph is then just the serial order of $\mathcal{S}$. From this we can calculate:

$$wl(\mathbb{Z}_n) = 2(n-1),$$

$$wl(\maltese_d) = \binom{2d+1}{3} - d^2,$$

(note that $wl(K_n) = \binom{n+1}{3}$ and that $\maltese_d$ is a complete graph on $2d$ vertices minus the edges between antipodal vertices) and

$$wl(Q_2) = 6,$$

as shown in Chapter 1.

(2) The 3-cube, Q_3: From the stability order of the cube, given in Fig. 3.4, its derived network is in Fig. 3.6. Each vertex in the Hasse diagram of Fig. 3.6 represents the set of 3-tuples to the left of the vertices which precede it. The number on the right of the vertex is the weight of that set, $|\Theta(S)|$. There is just one optimal s-t path in this derived network. It corresponds to lexicographic order of the vertices and $wl(Q_3) = 28$.

Exercise 3.3 *From the stability order of the 4-cube, shown in Fig. 3.5, diagram its derived network and show that there is just one optimal s-t path in this derived network. Give the corresponding optimal numbering. Hint: Build up stable sets systematically, one element at a time, starting with* $\emptyset$.

3.2.10 Summary

We have introduced the notion of stabilization for the wirelength problem and demonstrated its effectiveness by calculating the wirelength of $\mathbb{Z}_n$, $\maltese_d$, Q_3 and Q_4. The work required to calculate $wl(Q_d)$, $d > 4$, appears to be increasing rapidly, but the results for $d = 3, 4$ suggest that lexicographic order plays a special role for this problem. We shall continue, in Chapter 5, with further applications and extensions of stabilization.

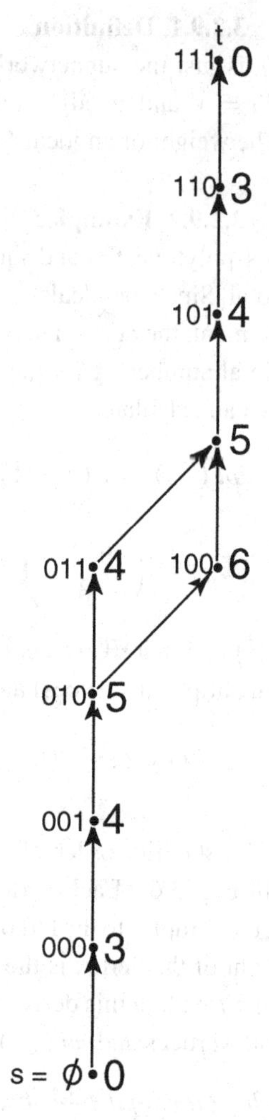

Fig. 3.6 Derived network for $\mathcal{S}(Q_3)$.

3.3 Compression

3.3.1 Introduction

Now we show how Steiner operations can be derived from direct product decompositions of a graph, G.

In Chapter 1 we demonstrated (Corollary 1.2) that

$$wl(G) \geq \sum_{k=0}^{n} \min_{|S|=k} |\Theta(S)| .$$

We also observed that the problem of minimizing $|\Theta(S)|$ over all k-sets of V is a combinatorial analog of the classical isoperimetric problem of Greek geometry. In Q_d the structure of the sets minimizing $|\Theta(S)|$ for $|S| = k$, the cubal sets, was crucial in proving their optimality. We would like to extend this proof to other graphs but to do so we must determine what about the cubal sets was essential for that proof (of Theorem 1.1) to work. Examining the proof, we see that it is only necessary to have one family of nested solutions, such as that given by the lexicographic numbering of the vertices of Q_d in order for the main simplifying step to work. This leads us to make the following definition. If G has a family of sets of vertices, $S_0 \subset S_1 \subset ... \subset S_n$, with $|S_k| = k$ and $|\Theta(S_k)| = \min_{|S|=k} |\Theta(S)|$, then it is said to have *nested solutions* for the EIP.

Lemma 3.2 *G has nested solutions if and only if $\exists$ a numbering $\eta : V \to \{1, 2, ..., n\}$ (one-to-one and onto), whose initial segments, $S_k(\eta)$, $k = 0, 1, ..., n$ are all solutions.*

If a graph has nested solutions then it greatly simplifies the problem of finding them. Starting with the null set, the unique solution for $k = 0$, we may assume that we have all solutions, S_k, which lie in some nested family, for $k \geq 0$. Then (under the hypothesis of nested solutions), we obtain all those for $k + 1$ by minimizing the *marginal boundary*

$$|\Theta(S_k + \{v\})| - |\Theta(S_k)|,$$

over all $v \notin S_k$. Note that this process may be further simplified by considering only stable sets, i.e. if a graph has nested solutions $S_0 \subset S_1 \subset ... \subset S_n$, then by Theorem 3.2, $Stab_\infty(S_0) \subset Stab_\infty(S_1) \subset ... \subset Stab_\infty(S_n)$ are also nested solutions.

3.3.2 Definition

If $G = H \times J$, H having nested solutions with a corresponding optimal numbering, η, recall that

$$\begin{aligned} V_{H\times J} &= V_H \times V_J = \{(v, w) : v \in V_H \text{ and } w \in V_J\}, \\ E_{H\times J} &= (E_H \times V_J) + (V_H \times E_J), \\ \partial_{H\times J} &= (\partial_H \times V_J) + (V_H \times \partial_J). \end{aligned}$$

For a set $S \subseteq V_{H\times J}$ let

$$Comp(S) = \bigcup_{w\in V_J} \left[S_{k_w}(\eta) \times \{w\}\right],$$

where $k_w = |S \cap (V_H \times \{w\})|$. In words, if G is viewed as being made up of copies of H, one for each vertex of J, then *the compression of* S is the set in which the intersection of S with each copy of H is replaced by the initial segment of η of that same size.

3.3.3 Basic properties of compression

Theorem 3.4 *For all* $S, T \subseteq V$,

(1) $|Comp(S)| = |S|$,
(2) $|\Theta(Comp(S))| \leq |\Theta(S)|$,
(3) $S \subseteq T$ *implies* $Comp(S) \subseteq Comp(T)$.

Proof

$$\begin{aligned}(1)\ |Comp(S)| &= \left|\bigcup_{w\in V_J}\left[S_{\ell_w}(\eta)\times\{w\}\right]\right| \\ &= \sum_{w\in V_J}\left|S_{\ell_w}(\eta)\times\{w\}\right| \\ &= \sum_{w\in V_J}|S\cap(V_H\times\{w\})| = |S|.\end{aligned}$$

(2) Note that edges connect corresponding vertices in two copies of H, say $H \times \{w_1\}$ and $H \times \{w_2\}$, if and only if w_1, w_2 are connected by an edge in J. Thus, for $S \subseteq V_{H\times J}$

$$\begin{aligned}&|\Theta(Comp(S))| \\ &\quad= \sum_{w\in V_J}\left|\Theta\left(S_{k_w}(\eta)\right)\right| + \sum_{\substack{V_J \\ f\in E_J \\ \partial(f)=\{w_1,w_2\}}}\left|k_{w_1}-k_{w_2}\right| \\ &\quad\leq \sum_{w\in V_J}|\Theta(S\cap(V_H\times\{w\}))| \\ &\qquad+ \sum_{\substack{f\in E_J \\ \partial(f)=\{w_1,w_2\}}}|\{v\in V_H : (v,w_1)\in S \text{ and } (v,w_2)\notin S\}| \\ &\qquad+ \sum_{\substack{f\in E_J \\ \partial(f)=\{w_1,w_2\}}}|\{v\in V_H : (v,w_1)\notin S \text{ and } (v,w_2)\in S\}| \\ &\quad= |\Theta(S)|.\end{aligned}$$

(3) If $S \subseteq T$ then

$$k_w = |S \cap (V_H \times \{w\})| \leq |T \cap (V_H \times \{w\})| = k'_w.$$

Therefore

$$\begin{aligned} Comp(S) &= \bigcup_{w \in V_J} \left[S_{k_w}(\eta) \times \{w\} \right] \\ &\subseteq \bigcup_{w \in V_J} \left[S_{k_w}(\eta) \times \{w\} \right] \\ &= Comp(T). \end{aligned}$$

□

Corollary 3.2 *Comp is a Steiner operation.*

In the proof of Theorem 1.1 compression was used only once. However, the graph of the d-cube, being a d-fold product of a single edge, has many factorizations as a product of lower-dimensional cubes. Can the compression operations derived from all of these factorizations be combined into one Steiner operation, $Comp^{(\infty)}$? And if so, can we compute its range in a simple, systematic way? As for stabilization, the answer is, with some small reservations, yes!

3.3.4 The compressibility order

Suppose the graph, G, is factorable as (i.e. is isomorphic to) the products of subgraphs, $H_1 \times J_1, H_2 \times J_2, ..., H_k \times J_k$. If $H_1, H_2, ..., H_k$ have nested solutions with numberings $\eta_1, \eta_2, ..., \eta_k$, respectively, then by the previous section they induce Steiner operations $Comp_1, Comp_2, ..., Comp_k$ on G. As we did to produce a "pushout" for repeated stabilizations, we define Steiner operations on G by composing the $Comp_i$ cyclically:

$$Comp^{(0)} \text{ is the identity on } N(G),$$

$$Comp^{(n)} = Comp_{n (\text{mod}\, k)} \circ Comp^{(n-1)}, \text{ for } n > 0.$$

A numbering, η of G, is called *consistent with* $\eta_1, \eta_2, ..., \eta_k$ if

(1) for all $w \in V_{J_i}$ the relative order of $\eta | V_{H_i} \times \{w\}$ is independent of w,
(2) the numbering of that common relative order is η_i.

Theorem 3.5 *If there is a numbering, η, of G, which is consistent with (the optimal numberings) $\eta_1, \eta_2, ..., \eta_k$, then the sequence $Comp^{(0)}$, $Comp^{(1)}$, $Comp^{(2)}$, ... is eventually constant.*

Proof For $S \subseteq V_G$, the sum $\eta(S) = \sum_{v \in V_G} \eta(v)$ is a positive integer and $\eta\left(Comp^{(i)}(S)\right) \leq \eta(S)$. Also $\eta\left(Comp^{(i)}(S)\right) = \eta(S)$ if and only if $Comp^{(i)}(S) = S$. Since $\eta\left(Comp^{(i)}(S)\right)$ is integer valued and cannot decrease indefinitely, it will be constant for n sufficiently large ($n > (k-1)$ $\eta(V)$). □

Denote this limit Steiner operation by $Comp^{(\infty)}$. A set $S \subseteq V_G$ will be called *compressed with respect to the compression operation Comp* if $Comp(S) = S$. A set $S \subseteq V_G$ is then in the range of $Comp^{(\infty)}$ if and only if it is compressed with respect to $Comp_i$, $1 \leq i \leq k$.

3.3.4.1 Definition

For each factorization, $G \simeq H_i \times J_i$, we have a partial order on V_G. It is the disjoint union of the total orders given by η_i on each copy, $H_i \times \{w\}$, of H_i. The *compressibility order*, $\mathcal{C}$, is then the reflexive and transitive closure of all these partial orders. That is if we let

$$P^{(0)} = \{(v, v) : v \in V_G\}, \text{ the identity on } V_G,$$

$$P^{(1)} = \bigcup_{i=1}^{k} \{(v, v') \in V_G \times V_G : v, v' \in H_i \times \{w\} \text{ for some } w \in J_i$$

$$\text{and } \eta_i(v) < \eta_i(v')\},$$

$$P^{(n+1)} = P^{(1)} \circ P^{(n)} \text{ for } n > 0$$

then

$$\mathcal{C} = \bigcup_{n=0}^{\infty} P^{(n)}.$$

3.3.4.2 Example

Let G_1 and G_2 be any graphs having nested solutions with optimal numberings, η_1, η_2 respectively. Lexicographic order on $G_1 \times G_2$ is consistent with respect to η_1, η_2 and the resulting compressibility order is just the product order. Similarly, the compressibility order of the product of any number of graphs which have nested solutions is the product order.

Theorem 3.6 *A set $S \subseteq V_G$ is compressed with respect to $Comp_1$, $Comp_2$, ..., $Comp_k$ if and only if it is an ideal in the compressibility order, $\mathcal{C}$.*

Exercise 3.4 *Prove Theorem 3.6 (see the proof of Theorem 3.3).*

It follows then that the range of $Comp^{(\infty)}$ is $N(\mathcal{C})$, the derived network of the compressibility order.

3.3.5 Another solution of the wirelength problem for Q_d

With the machinery of compression, the solution of this challenging problem is now easy. We know that Q_1 and Q_2 have nested solutions and that lex(icographic) numbering is optimal, so, by way of induction, assume it for Q_{d-1}, $d > 2$. In calculating $\mathcal{C}_d$, there are many factorizations, $Q_d \simeq H \times J$, but we may restrict our attention to a small number of these. Clearly, H_i should be a maximal factor, so let J_i be the ith component. Then H_i consists of the product of the other $d-1$ components, making it a $(d-1)$-cube. Lex numbering of the d-cube, restricted to $H_i \times \{0\}$ or $H_i \times \{1\}$ is still lexicographic order, which by induction is optimal. Thus lex numbering is consistent. For any i, $1 \leq i \leq d$, the contribution of $H_i \times J_i$ to $\mathcal{C}_d$ is that the two subcubes, $H_i \times \{0\}$ and $H_i \times \{1\}$, are totally ordered (lexicographically). If $i = 1$, the maximum member of the half given by $x_1 = 0$ is 01^{d-1}, and the minimum member of the half given by $x_1 = 1$ is 10^{d-1}. Since these d-tuples have no common components, they are not directly comparable in $\mathcal{C}_d$. The vertex which covers 10^{d-1} in $H_d \times \{1\}$ is $10^{d-2}1$ and that does agree with 01^{d-1} in the dth coordinate so $01^{d-1} <_{\mathcal{C}_d} 10^{d-2}1$. Similarly, $01^{d-2}0 <_{\mathcal{C}_d} 10^{d-1}$. Thus 01^{d-1} and 10^{d-1} are incomparable and the only incomparables in $\mathcal{C}_d$. Its Hasse diagram and derived network are shown in Fig. 3.7

Since the path with $1^{d-1}0$ before $0^{d-1}1$ has lower weight, the best choice is clear. In fact, this inequality determines one last Steiner operation which takes $N(\mathcal{C}_d)$ to the single path given by lex numbering. The wirelength of Q_d is thus that of lex numbering. With respect to lex numbering, the 2^{d-1} edges determined by the ith coordinate contribute a difference of 2^i. Therefore

$$wl(Q_d) = 2^{d-1} \sum_{i=1}^{d} 2^i = 2^{d-1}\left(2^d - 1\right).$$

3.4 Comments

The concept of stabilization was abstracted from a paper by Bernstein, Steiglitz and Hopcroft [**14**]. Having asked the question of how symmetry might be used to systematically simplify combinatorial isoperimetric problems, the author searched the literature (*c.* 1975) and found just that one relevant paper. Bernstein *et al.* defined one-dimensional and two-dimensional stabilization for

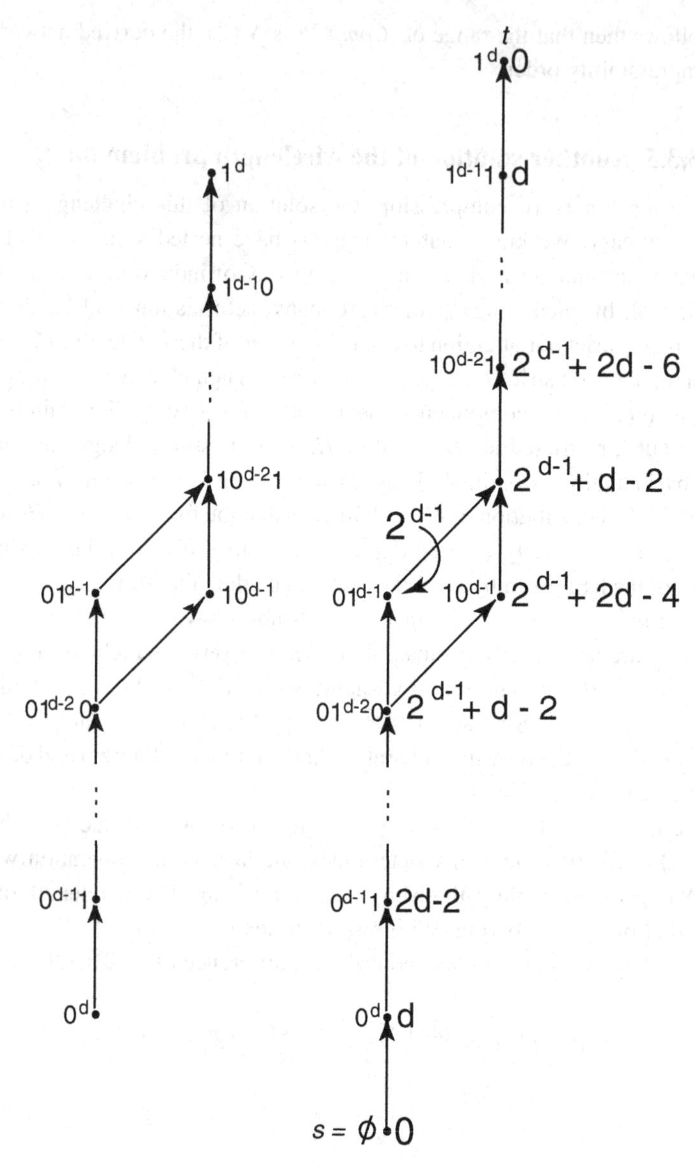

Fig. 3.7 $\mathcal{C}(Q_d)$ and $N(\mathcal{C}(Q_d))$.

Q_d but the underlying idea was the same for both, and extended to arbitrary graphs with stabilizing reflections. The general definition of stabilization, its properties, the "pushout" for multiple stabilizations, the stability order and derived network, were first presented in [**48**]. Although stabilization is based on reflective symmetry, it is much more powerful (when it works) than modding out the action of the symmetry group. One can see this already in the square, which has two inequivalent 2-sets and three inequivalent numberings but only one of each is stable. As the graphs get larger, the disparity becomes far greater. Not only is stabilization stronger (when it works), it is much easier to generate all stable sets (as ideals in the stability order) than all equivalence classes or representatives for them.

Compression, in contrast to stabilization, has appeared in most of the papers on combinatorial isoperimetric problems, independently discovered many times. It has been the dominant proof technique in the field. Our exposition, however, is the first to make the compressibility order explicit. The power of compression is derived from the fact that it uses the inductive hypothesis (factors have nested solutions) over and over. Kleitman was the first to observe that the systematic application of compression could simplify the proof of the EIP on Q_d, leaving the relative order of just two elements to be settled.

4
The vertex-isoperimetric problem

Having developed the Steiner operations, stabilization and compression, for the edge-isoperimetric problem, we now explore the possibility of applying them to another isoperimetric problem on graphs, the vertex-isoperimetric problem (VIP). This chapter revisits the material of the first three chapters with the VIP in place of the EIP. The development of global methods for the VIP can be condensed because it is largely the same, but there are differences and the differences are instructive.

4.1 Definitions and examples

4.1.1 The VIP

For a graph $G = (V, E, \partial)$ and $S \subseteq V$, the *vertex-boundary* of S is

$$\Phi(S) = \{w \in V - S : \exists e \in E,\ \partial(e) = \{v, w\} \text{ and } v \in S\}.$$

In words, the vertex boundary of S is the set of vertices not in S but having neighbors in S. Then given $k \in \mathbb{Z}^+$, *the vertex-isoperimetric problem* is to minimize $|\Phi(S)|$ over all $S \subseteq V$ such that $|S| = k$. The VIP is trivial on K_n so we look at $\mathbb{Z}_n$. First, we exhibit, in Fig. 4.1, some 2-sets in $\mathbb{Z}_8$ which have vertex-boundaries of differing sizes.

4.1.1.1 Exercise

Show that $\forall n$ and $0 < k < n$,

$$\min_{\substack{S \subseteq \mathbb{Z}_n \\ |S|=k}} |\Phi(S)| = 2.$$

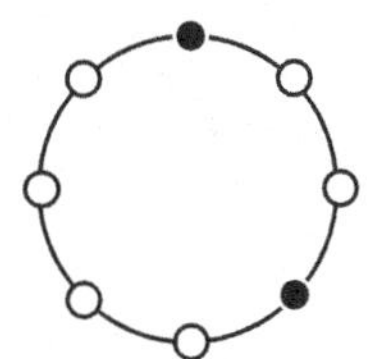
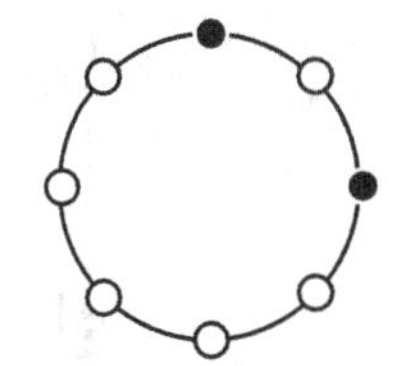
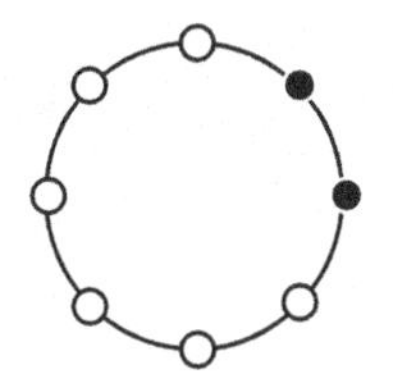

Fig. 4.1 $|\Phi(S)| = 4, 3, 2$ respectively.

4.2 Stabilization and VIP

Theorem 4.1 *$\forall S \subseteq V$, $\left|\Phi\left(Stab_{\mathcal{R},p}(S)\right)\right| \leq |\Phi(S)|$.*

Proof Suppose that $w \in \Phi\left(Stab_{\mathcal{R},p}(S)\right)$ but $w \notin \Phi(S)$. Clearly $w \neq \mathcal{R}(w)$ so there are two possibilities:

(1) $\|w - p\| < \|\mathcal{R}(w) - p\|$: $\exists e \in E,\ \partial(e) = \{v, w\}$ and $v \in Stab_{\mathcal{R},p}(S)$. But since $w \notin \Phi(S)$, $v \notin S$. Therefore $\mathcal{R}(v) \in S$. Now by the definition of stabilization, $w \notin Stab_{\mathcal{R},p}(S)$ which implies that $\mathcal{R}(w) \notin S$. Since $\partial(\mathcal{R}(e)) = \{\mathcal{R}(v), \mathcal{R}(w)\}$, $\mathcal{R}(w) \in \Phi(S)$. Also by a similar argument $\mathcal{R}(w) \notin \Phi\left(Stab_{\mathcal{R},p}(S)\right)$. Thus if $w \in \Phi\left(Stab_{\mathcal{R},p}(S)\right)$ but $w \notin \Phi(S)$ then $\mathcal{R}(w) \in \Phi(S)$ but $\mathcal{R}(w) \notin \Phi\left(Stab_{\mathcal{R},p}(S)\right)$ which balances it out.

(2) $\|\mathcal{R}(w) - p\| < \|w - p\|$:

□

Exercise 4.1 *Complete the proof (Case 2).*

Corollary 4.1 *Stabilization is a Steiner operation for VIP.*

That is to say $Stab_{\mathcal{R},p}$, acting on the derived network $N(G)$ weighted by $|\Phi(S)|$ rather than $|\Theta(S)|$, is a pathmorphism. This follows from Theorem 3.2, with part (2) replaced by Theorem 4.1, and Theorem 2.3. The construction of the "pushout" for multiple stabilizations for VIP is essentially the same as it was for EIP, only the weights on stable sets being different.

Let us now apply stabilization to the VIP on Q_d. For $d = 2$, as we saw before, there is only one stable set of each cardinality, so they are solutions of the VIP by default. For $d = 3$ we have the stability order of Q_3 (Fig. 3.4) and its Φ-weighted derived network in Fig. 4.2 (compare it to Fig. 3.6). We see then that the following table gives the solution of the VIP on Q_3

k	0	1	2	3	4	5	6	7	8
$\min_{S \subseteq \mathbf{Q}_3, \lvert S\rvert = k} \lvert\Phi(S)\rvert$	0	3	4	4	3	3	2	1	0

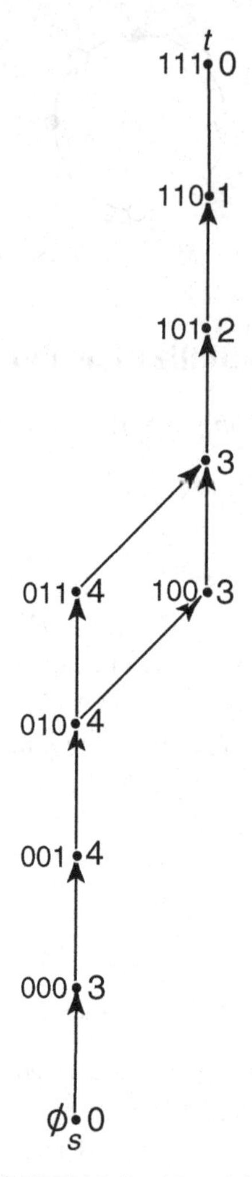

Fig. 4.2 $N(\mathcal{S}(Q_3))$ with weight $|\Phi|$.

and that there is one stable numbering (path in the derived network of the stability order) whose initial segments achieve those minima. That numbering is not lex but only differs from lex by an interchange of two vertices.

For $d = 4$ we have the stability order of Q_4 (Fig. 3.5) and its Φ-weighted derived network in Fig. 4.3.

We have simplified the Hasse diagram of Fig. 4.3, leaving the arrows off, and will continue to do so. By convention, all edges are directed upward.

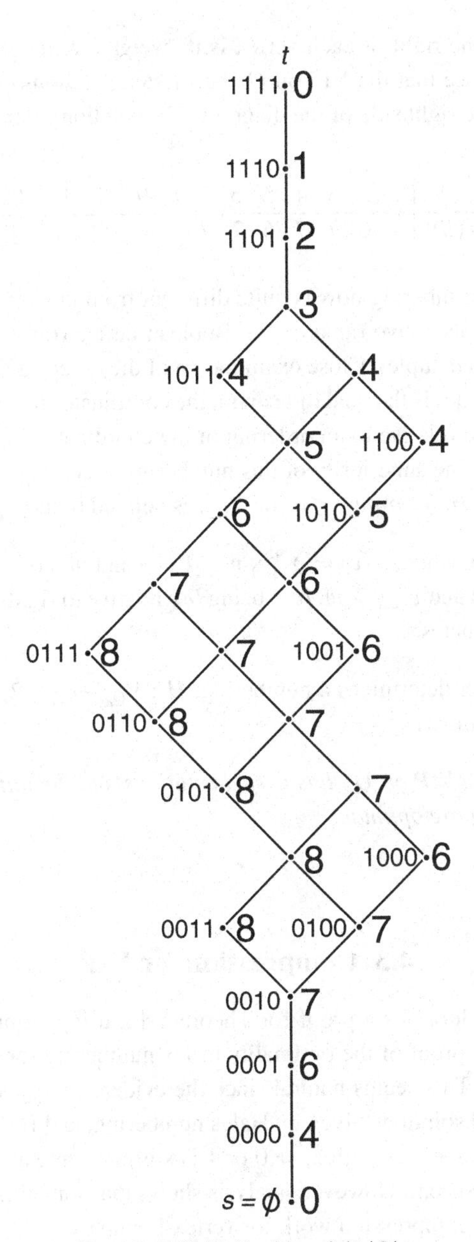

Fig. 4.3 $N(\mathcal{S}(Q_4))$ with weight $|\Phi|$.

The number to the right of each vertex is its weight (value of $|\Phi(S)|$). From Fig. 4.3 we can see that the VIP on Q_4 has nested solutions corresponding to an s-t path up the right side of the diagram. The solution is summarized in the table

k	1	2	3	4	5	6	7	8	9	10	11	12	13	14	15
$\min_{\substack{S\subseteq Q_4 \\ \lvert S\rvert=k}} \lvert\Phi(S)\rvert$	4	6	7	7	6	7	7	6	6	5	4	4	3	2	1

The optimal numbering now is quite different from lex numbering. It takes the vertices by successive ranks in the Boolean lattice (rank in $\mathcal{B}_d$ being the number of 1s in a d-tuple). Close examination of the ordering within each rank in $\mathcal{B}_d$ suggests that it is the dual of lex with the coordinates reversed (remember that lex is defined relative to an ordering of the coordinates, which we took to be left-to-right). The superiority of this numbering was first pointed out by A. Hales (see [**46**]). *Hales order*, $\leq_H$, on V_{Q_d}, is defined by $v \leq_H w$ if

(1) $r(v) < r(w)$, where $r(v) = \sum_{i=1}^{d} v_i$ is Boolean rank, or
(2) $r(v) = r(w)$ and $v \geq_{lex'} w$, lex' being lex relative to right-to-left ordering of the coordinates.

This total order determines a numbering, $H : V_{Q_d} \to \{1, 2, ..., 2^d\}$, which we call Hales numbering

Theorem 4.2 *The VIP on Q_d has nested solutions and the initial segments of Hales numbering are optimal.*

4.3 Compression for VIP

Suppose that we look for a proof for Theorem 4.2 using compression, along the lines of our reproof of the optimality of lex numbering for edge-boundary in Section 3.3.5. This seems natural since the evidence suggests that the VIP on Q_d has nested solutions given by Hales numbering, and Hales order on Q_d restricted to the subcubes with $x_i = 0$ or 1 is still the same as Hales order on Q_{d-1} so it is consistent. However, analysis shows that something additional is required to make compression work for vertex-boundary.

A nested family of sets, $S_0 \subset S_1 \subset ... \subset V$ with $|S_k| = k$, is called *generative* if $\forall k$

$$S_k + \Phi(S_k) = S_{k+|\Phi(S_k)|}.$$

Note that any generative family of nested sets contains all the balls (in the path-metric on G) centered at the unique vertex $v \in S_1$.

If $G = H \times J$ and H has nested solutions (for the VIP) with optimal numbering η whose initial segments are generative, we define the compression operation, *Comp* (S), in exactly the same way as we did in Definition 3.3.2 of Chapter 3: For a set $S \subseteq V_{H \times J}$

$$Comp\,(S) = \bigcup_{w \in V_J} \left[S_{k_w}(\eta) \times \{w\} \right].$$

Then we have

Theorem 4.3 *For all $S, T \subseteq V$,*

(1) $|Comp\,(S)| = |S|$,
(2) $|\Phi\,(Comp\,(S))| \leq |\Phi\,(S)|$,
(3) *$S \subseteq T$ implies $Comp\,(S) \subseteq Comp\,(T)$.*

Proof The only difference between this statement and Theorem 3.4 is that Φ has replaced Θ, so we need only address part (2). Note that edges connect corresponding vertices in two copies of H, say $H \times \{w_1\}$ and $H \times \{w_2\}$, if and only if w_1, w_2 are connected by an edge in J. Thus, for $S \subseteq V_{H \times J}$

$$\begin{aligned}
|\Phi\,(Comp\,(S))| &= \sum_{w_1 \in V_J} \left[\max \left\{ \left|\Phi\left(S_{k_{w_1}}(\eta)\right)\right| \right\} \cup \left\{ k_{w_2} - k_{w_1} : \right.\right. \\
&\qquad \exists f \in E_J,\ \partial(f) = \{w_1, w_2\}\}] \\
&\leq \sum_{w_1 \in V_J} |\Phi\,(S \cap (V_H \times \{w_1\})) \cup \{(v, w_1) \notin S : \\
&\qquad \exists f \in E_J,\ \partial(f) = \{w_1, w_2\},\ (v, w_2) \in S\}| \\
&= |\Phi\,(S)|.
\end{aligned}$$

□

Corollary 4.2 *Compression (with respect to a generative family of nested solutions) is a Steiner operation for vertex-boundary.*

The remainder of the theory of compression for Φ is the same as that developed for Θ in Chapter 3. If G is factorable as products of subgraphs, $H_1 \times J_1$, $H_2 \times J_2$, ..., $H_k \times J_k$, such that $H_1, H_2, ..., H_k$ have nested solutions with generative optimal numberings $\eta_1, \eta_2, ..., \eta_k$, respectively, then they induce

Steiner operations $Comp_1, Comp_2, ..., Comp_k$ on G. We then define Steiner operations on G by composing the $Comp_i$ cyclically:

$$Comp^{(0)} \text{ is the identity on } N(G),$$

$$Comp^{(n)} = Comp_{n(\text{mod}\, k)} \circ Comp^{(n-1)}, \text{ for } n > 0,$$

and we have

Theorem 4.4 *If there is a numbering, η, of G, which is consistent with (the generative optimal numberings) $\eta_1, \eta_2, ..., \eta_k$, then the sequence $Comp^{(0)}, Comp^{(1)}, Comp^{(2)}, ...$ is eventually constant.*

4.3.1 Compressibility order

For each factorization, $G \simeq H_i \times J_i$, we have a partial order on V_G. It is the disjoint union of the total orders given by η_i on each copy, $H_i \times \{w\}$, of H_i. The *compressibility order*, $\mathcal{C}$, is then the symmetric and transitive closure of all these partial orders. That is we let

$$P^{(0)} = \{(v, v) : v \in V_G\}, \text{ the identity on } V_G,$$

$$P^{(1)} = \bigcup_{i=1}^{k} \{(v, v') \in V_G \times V_G : v, v' \in H_i \times \{w\} \text{ for some } w \in J_i$$
$$\text{and } \eta(v) < \eta(v')\},$$

$$P^{(n+1)} = P^{(1)} \circ P^{(n)} \text{ for } n > 0,$$

and then

$$\mathcal{C} = \bigcup_{n=0}^{\infty} P^{(n)}.$$

Theorem 4.5 *A set $S \subseteq V_G$ is compressed with respect to $Comp_1, Comp_2, ..., Comp_k$ if and only if it is an ideal in the compressibility order, $\mathcal{C}$.*

Again, the range of $Comp^{(\infty)}$ is $N(\mathcal{C})$, the derived network of the compressibility order. The only difference from Chapter 3 is that $N(\mathcal{C})$ now has weights $|\Phi(S)|$ rather than $|\Theta(S)|$.

4.4 Optimality of Hales numbering

Proof (of Theorem 4.2). By induction on dimension, d. We have already proved it for $d \leq 4$, so assume the theorem holds for $d - 1 \leq 4$ and apply compression to the d-cube. The Hales numbering on Q_d is consistent with that on all $(d-1)$-cubes ($v_i = 0$ or 1) so a compressibility order is defined. Let us calculate it. Note that

(1) If for $v, w \in V_{Q_d}$ there exists an i such that $v_i = w_i$, then v and w are comparable in the compressibility order.
(2) So if v and w are not comparable in $\mathcal{C}$, then $w = \overline{v}$ where

$$\overline{v}_i = \begin{cases} 1 & \text{if } v_i = 0 \\ 0 & \text{if } v_i = 1. \end{cases}$$

(3) As before, $\mathcal{B}_d$, the coordinatewise partial order on V_{Q_d}, is a suborder of $\mathcal{C}$.

Let $r(v) = \sum_{i=1}^{d} v_i$, the rank of v in $\mathcal{B}_d$ and $\mathcal{B}_{d,k} = \{v \in V : r(v) = k\}$. Now look at any $v, w \in V_{Q_d}$, $v \neq w$. If $r(v), r(w) < d/2$ then, by the pigeonhole principle, there is an i such that $v_i = 0 = w_i$ so v and w must be comparable in $\mathcal{C}$. Similarly, if $r(v), r(w) > d/2$ there is an i such that $v_i = 1 = w_i$ so v and w are comparable in $\mathcal{C}$. If d is odd then

$$\bigcup_{k<d/2} \mathcal{B}_{d,k}$$

is totally ordered by $\mathcal{C}$ as is

$$\bigcup_{k>d/2} \mathcal{B}_{d,k}.$$

The maximum element of the lower half is $1^{(d-1)/2}0^{(d+1)/2}$ and the minimum element of the upper half is $0^{(d-1)/2}1^{(d+1)/2}$. They are not directly comparable but the successor of $0^{(d-1)/2}1^{(d+1)/2}$ is $0^{(d-3)/2}101^{(d-1)/2}$ which is also above $1^{(d-1)/2}0^{(d+1)/2}$ since they coincide in several components and the latter has greater rank. By the same reasoning, the predecessor of $1^{(d-1)/2}0^{(d+1)/2}$ is $1^{(d-3)/2}010^{(d-1)/2}$ which is below $0^{(d-1)/2}1^{(d+1)/2}$. Thus there are just those two incomparables in the compressibility order.

If d is even,

$$\bigcup_{k<d/2} \mathcal{B}_{d,k} \cup \left(\mathcal{B}_{d,d/2} \cap \{v \in V : v_d = 1\}\right)$$

is totally ordered by $\mathcal{C}$ as is

$$\bigcup_{k>d/2} \mathcal{B}_{d,k} \cup \left(\mathcal{B}_{d,d/2} \cap \{v \in V : v_d = 0\}\right).$$

The maximum element of the lower half is $1^{d/2-1}0^{d/2}1$ and the minimum element of the upper half is $0^{d/2-1}1^{d/2}0$.

Exercise 4.2 *Show that* $1^{d/2-1}0^{d/2}1$ *and* $0^{d/2-1}1^{d/2}0$ *are the only incomparables in* $\mathcal{C}$.

All that remains then in showing that Hales numbering is optimal, is to show that the compressed set of cardinality 2^{d-1} which contains $1^{(d-1)/2}0^{(d+1)/2}$, if d is odd, or $1^{d/2-1}0^{d/2}1$, if d is even, has a value of $|\Phi|$ no greater than that which contains $0^{(d-1)/2}1^{(d+1)/2}$ or $0^{d/2-1}1^{d/2}0$. The easy way to do this is to look at the marginal contribution of the two elements, but let us extend the idea a bit, in a way which will be useful later.

Lemma 4.1 *If* $S \subseteq V_{Q_d}$ *is stable,* $S \neq \emptyset$*, and* $a \in V_{Q_d} - S$ *is minimal (with respect to stability order), then* $S + \{a\}$ *is stable and*

$$|\Phi(S + \{a\})| - |\Phi(S)| = j_0(a)$$

where

$$j_0(a) = \begin{cases} d & \text{if } a = 0^d, \\ \min\{j : a_j = 1\} - 2 & \text{otherwise.} \end{cases}$$

Proof (of the lemma). Consider $a + \delta_j \in V_{Q_d}$. If $j < j_0$, then $a + \delta_j \in \Phi(S + \{a\})$ but $a + \delta_j \notin \Phi(S)$ since $a + \delta_j - \delta_i \in S$ implies that $i \geq j_0 > j$ which implies that $a <_S a + \delta_j - \delta_i$ which implies that $a \in S$, a contradiction. On the other hand, if $j > j_0$, then $a + \delta_j \in \Phi(S + \{a\})$ but also $a + \delta_j \in \Phi(S)$ since $a + \delta_j - \delta_{j_0} <_S a$ so that $a + \delta_j - \delta_{j_0} \in S$ by the minimality of a in $V_{Q_d} - S$. □

Applying this lemma to the derived network, $N(\mathcal{C}(Q_d))$ above, completes the proof of Theorem 4.2. □

4.5 Applications to layout problems

4.5.1 The bandwidth problem

On a graph, $G = (V, E, \partial)$, the *bandwidth problem* is to minimize

$$bw(\eta) = \max_{\substack{e \in E \\ \partial(e)=\{v,w\}}} |\eta(v) - \eta(w)|$$

over all numberings, η, of G. We then define *the bandwidth of* G to be

$$bw(G) = \min_{\eta} bw(\eta)$$

This problem originated in numerical analysis, where the edges of G represent nonzero entries in an $n \times n$ matrix and η a permutation of its rows and columns. $bw(\eta)$ then represents the width of a band (about the main diagonal) which contains all nonzero entries of the matrix (whose rows and columns have been reordered according to η. The smaller its bandwidth is, the more efficient numerical calculations with the matrix will be, so the objective is to minimize it. The same problem may also be interpreted as a layout problem, laying out the graph (wiring diagram of an electronic circuit) on a linear chassis so as to minimize the maximum length of any wire.

4.5.1.1 Example

On the square we recall there are three nonisomorphic numberings. These were shown in Fig. 1.3 and repeated here in Fig. 4.4. Only the first minimizes the bandwidth.

Exercise 4.3 *Show that* $bw(\mathbb{Z}_n) = 2$.

Exercise 4.4 *Show that* $bw(K_n) = n - 1$.

For more background on the bandwidth problem, see the survey by Chinn *et al.* [**26**].

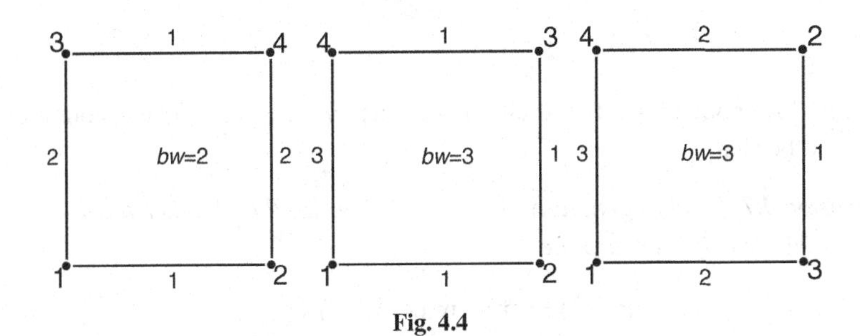

Fig. 4.4

4.5.2 Reducing bandwidth to minimum path

Given a graph, G, for the bandwidth problem, we construct its derived network, $N(G)$, with weights $|\Phi(S)|$. We let the weight of an s-t path, P, be

$$|\Phi|(P) = \max_{S \in P} |\Phi(S)|.$$

By the remark of Klee in Section 2.2.5,

$$\min_{P \in \mathfrak{P}(N)} |\Phi|(P),$$

the minimum weight of any s-t path in $N(G)$, may be computed by a variant of the acyclic algorithm. Then we have

Theorem 4.6 *The minimum path problem for $N(G)$ gives a lower bound for the bandwidth of G, i.e.*

$$bw(G) \geq \min_{\eta} \max_{0 \leq k \leq n} |\Phi(S_k(\eta))|.$$

Proof For each numbering, η, and each k, $1 \leq k \leq n$, there exists $v \in V$ such that $\eta(v) = k$. Then

$$\max_{\substack{e \in E \\ \partial(e) = \{u,w\}}} |\eta(u) - \eta(w)| \geq |\Phi(S_k(\eta))|,$$

since some vertex, $w \in \Phi(S_k(\eta))$, must have $\eta(w) \geq k + |\Phi(S_k(\eta))|$ and $\exists e \in E$, $\partial(e) = \{u, w\}$, $u \in S_k(\eta)$ so $\eta(u) \leq k$ and then $|\eta(u) - \eta(w)| \geq (k + |\Phi(S_k(\eta))|) - k = |\Phi(S_k(\eta))|$. Therefore

$$bw(\eta) \geq \max_{0 \leq k \leq n} |\Phi(S_k(\eta))|$$

and

$$bw(G) \geq \min_{\eta} \max_{0 \leq k \leq n} |\Phi(S_k(\eta))|.$$

□

This lower bound is not generally sharp, but there is a nice sufficient condition for it to be sharp.

Theorem 4.7 *If G has generative nested solutions for the VIP, then the inequality of Theorem 4.6 is sharp, i.e.*

$$bw(G) = \min_{\eta} \max_{0 \leq k \leq n} |\Phi(S_k(\eta))|.$$

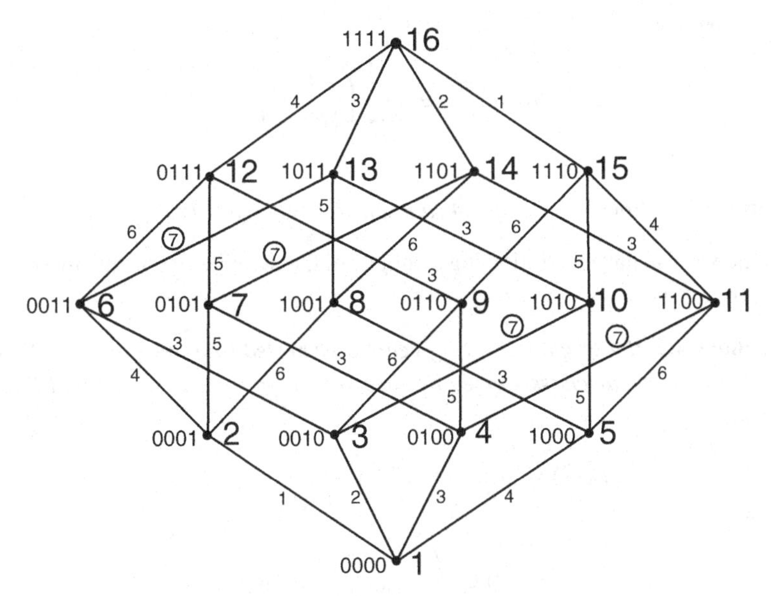

Fig. 4.5 Hales numbering of Q_4.

Proof In the proof of Theorem 4.6, take η_0 to be an optimal numbering and $e_0 \in E$, $\partial(e_0) = \{v_0, w_0\}$ to be an edge such that $\eta_0(v_0) < \eta_0(w_0)$ and $\eta_0(w_0) - \eta_0(v_0) = bw(G)$. Then $\eta_0(w_0) \leq \eta_0(v_0) + \left|\Phi\left(S_{\eta_0(v_0)}(\eta_0)\right)\right|$ so $\eta_0(w_0) = \eta_0(v_0) + \left|\Phi\left(S_{\eta_0(v_0)}(\eta_0)\right)\right|$. If there was a $u \in V$ with $\eta_0(u) < \eta_0(v_0)$ connected to w_0, it would contradict the definition of e_0. □

Corollary 4.3 *Hales numbering, H, minimizes bandwidth on the d-cube, i.e.*

$$bw(H) = bw(Q_d).$$

In Fig. 4.5 the small numbers on edges are the absolute differences. Maximum values are circled. Note that Hales numbering is not superior to lex on Q_3 but it is on Q_4.

Looking back at our solutions of the EIP on Q_2, Q_3 and Q_4 we have the following table of solutions for their bandwidth problems.

d	1	2	3	4
$bw(Q_d)$	1	2	4	7

In general we have the formula

Corollary 4.4

$$bw(Q_d) = \sum_{k=0}^{d-1} \binom{k}{\lfloor k/2 \rfloor}$$

Exercise 4.5 *Prove Corollary 4.4 (it is surprisingly difficult).*

And we also have the following analogue of the Steiglitz–Bernstein theorem (Theorem 4.1)

Corollary 4.5 *If a graph G, having generative nested solutions for the VIP, is laid out on a linear chassis given by sites $s_1 < s_2 < \ldots < s_n$ on the real line, $\mathbb{R}$, then*

$$wl(G) = \min_{\eta} \max_{\substack{e \in E \\ \partial(e)=\{v,w\}}} \left| s_{\eta(v)} - s_{\eta(w)} \right|$$

$$= \max_{0<k<n} \left(s_{k+\min_{S \subseteq V, |S|=k} |\Phi(S)|} - s_k \right).$$

Conversely, any graph whose solution of this generalized wirelength problem is of this form for all $s_1 < s_2 < \ldots < s_n$ must have generative nested solutions for the VIP.

Exercise 4.6 *Prove Corollary 4.5.*

4.5.3 Partitioning to minimize pins

In Section 1.3.3 of Chapter 1 we considered the problem of laying out a wiring diagram on p chips so as to minimize the number of wires which connect components on different chips. This is equivalent to p-partitioning the vertices of a graph, G, into essentially equal sized sets, so as to minimize the number of edges incident to vertices in different blocks of the partition. We observed that for many values of p it is possible to uniformly partition Q_d into p cubal sets which must then minimize the total number of connecting edges. A closely related layout problem is to uniformly p-partition the graph so as to minimize the total number of vertices which have neighbors in different blocks of the partition. Such a vertex requires a "pin" to be placed on the chip to which wires can be connected, so we are minimizing pins.

Our $\Phi(S)$ might be called the *external vertex-boundary of S* because it consists of vertices *not* in S which are connected to vertices in S. If S is a block of the partition, then the vertices which give pins are those of $\Phi(V - S)$, the external vertex-boundary of the complement of S or equivalently, the *internal vertex-boundary of* S. Thus the total number of pins for a partition, π, is

$$\sum_{S \in \pi} |\Phi(V - S)|.$$

It is not true, however, that $|\Phi(V - S)| = |\Phi(S)|$. On Q_d we showed that the initial segments of Hales numbering minimize $|\Phi(S)|$ for each $k = |S|$. The antipodal map (on Q_d), applied to Hales numbering gives the reverse of Hales numbering. The antipodal map is an isomorphism of Q_d so the terminal segments of Hales numbering also minimize $\Phi(V - S)$ which is the internal vertex-boundary of S. Thus initial segments of Hales numbering minimize the internal, as well as the external, vertex-boundary for their cardinality.

For $c \in Q_d$ and $r \in \mathbb{Z}^+$ let

$$B(c; r) = \{x \in Q_d : |x - c| \leq r\},$$

the *Hamming ball of radius* r *centered at* c. The Hamming balls $B\left(0^d; r\right)$, centered at the all-zero vector, are initial segments of Hales numbering, so they, and therefore all Hamming balls, minimize the internal vertex-boundary for their cardinality. In fact Hamming balls are the most desirable kind of blocks for a partition since they locally minimize $|\Phi(S)|$ (and $|\Phi(V - S)|$), as a function of $k = |S|$. This was made precise by Nigmatullin [**81**] who showed

Theorem 4.8 *If*

$$\sum_{i=0}^{r} \binom{d}{i} < k < \sum_{i=0}^{r+1} \binom{d}{i},$$

and $S \subset Q_d$, $|S| = k$, *then letting*

$$\alpha = \frac{k - \sum_{i=0}^{r} \binom{d}{i}}{\binom{d}{r+1}},$$

$$\beta = \frac{\sum_{i=0}^{r+1} \binom{d}{i} - k}{\binom{d}{r+1}}$$

we have $\alpha + \beta = 1$ *and*

$$|\Phi(S)| > \alpha \binom{d}{r+1} + \beta \binom{d}{r+2}.$$

So, if we can find a uniform partition of Q_d into Hamming balls it would give a particularly efficient solution to the pin-minimization problem. It so happens that this is a well-studied problem in algebraic coding theory [**82**]. It is known that the only uniform p-partitions of Q_d into Hamming balls of radius r are given in the table:

p	r	π
1	d	$\{Q_d\}$
2	$(d-1)/2$, d odd	$\{B(0^d;(d-1)/2), B(1^d;(d-1)/2)\}$
2^{12}	3	$\{B(x;3) : x \in G_{23}\}$, G_{23} the binary Golay code with $d = 23$
2^{2^n-n-1}	1	$\{B(x;1) : x \in H_d\}$, H_d the Hamming code with $d = 2^n - 1$
2^d	0	Q_d

Unfortunately, the partitions that are interesting in coding theory are not very interesting for our layout problem. The reason is illustrated by the Golay code, G_{23}: The size of a Hamming ball, $B(x;3)$, in $\{0,1\}^{23}$ is $\sum_{i=0}^{3}\binom{23}{i} = 1 + 23 + 253 + 1771 = 2048$, whereas $\left|\Phi\left(\{0,1\}^{23} - B(x;3)\right)\right| = \binom{23}{3} = 1771$. For large values of p, the internal vertex-boundary, $\binom{d}{r}$, of $B(x;r)$ is increasing exponentially with r so that it is almost as large as $|B(x;r)| = \sum_{i=0}^{r}\binom{d}{i}$. The 2-partition for d odd might be useful, but is not very interesting mathematically. There may also be uniform p-partitions for small p such as 3 or 4 with blocks which are near optimal. It has been suggested (see [**82**]) that simulated annealing will probably find such partitions.

4.6 Comments

This chapter broaches a fundamental question of global theories: having found a nice notion of morphism for a given problem, such as the Steiner operations, stabilization and compression, for the EIP, what other problems are preserved by those morphisms? What we have shown is that the VIP is also preserved by stabilization and compression.

Just as the EIP and wirelength problem have a symbiotic relationship, the VIP and bandwidth problem do also. Actually, it was those layout problems which led to the isoperimetric problems, but they also showed (through nested solutions for the EIP and generative nested solutions for the VIP) how to find the solutions and how to prove them. The isoperimetric problems, however, are clearly fundamental. They are relatively easy to prove, lead to many other applications and are analogs of the classical isoperimetric problem of Greek geometry, long recognized as fundamental.

The wirelength of a numbering (see Section 1.3.1),

$$wl(\eta) = \sum_{\substack{e \in E \\ \partial(e) = \{v, w\}}} |\eta(v) - \eta(w)|,$$

is what analysts would call an L_1 functional. Bandwidth,

$$bw(\eta) = \max_{\substack{e \in E \\ \partial(e) = \{v, w\}}} |\eta(v) - \eta(w)|$$

is L_∞ in that same sense. What about L_p, $1 < p < \infty$? The L_2 problem, minimizing

$$\sum_{\substack{e \in E \\ \partial(e) = \{v, w\}}} \left(\eta(v) - \eta(w)^2\right),$$

over all numberings, η, has been solved on Q_d by Crimmins *et al.* [**30**] using harmonic analysis on the dyadic group. The methods for these three cases are so different that a common generalization seems unlikely.

Spectral methods, based on the eigenvalues of certain matrices which are analogs of the Laplacian, have been very successful in bounding isoperimetric parameters of G and are widely applied. For more on this approach consult the monographs by Chung [**27**] and Lubotzky [**75**]. Our approach to the same problem, based on discrete analogs of Steiner symmetrization, is not as flexible as the spectral methods but does give better results in some important cases (see [**69**]).

The definition of compression for VIP originally appeared in [**79**], where it was used to solve a more general problem. We shall return to that result in Chapter 6.

5
Stronger stabilization

In this chapter we strengthen the theory of stabilization by:

(1) utilizing Coxeter's theory of groups generated by reflections to
   (a) facilitate the calculation of stability orders, and
   (b) generate a large family of graphs to which stabilization applies:
(2) weakening the requirement for a reflective symmetry of a graph to be stabilizing so that stabilization applies to a larger class of diagrams.

5.1 Graphs of regular solids

The edge-isoperimetric problem (EIP) and VIP are NP-complete in general but, as we have seen, a number of special cases, important in applications, have been solved and many of the solved cases are highly symmetric. In the early 1970s I became intrigued with the EIP on the graph of the dodecahedron. The idea was to take the combinatorial isoperimetric problems whose development had been motivated by applications to engineering, and rethink them from the viewpoint of pure mathematics. It is not difficult, assuming that the EIP on the dodecahedron has nested solutions, to find an optimal numbering, and that numbering has structure similar to lex on Q_d. The challenge was to prove that its initial segments are solution sets. Since the dodecahedron, unlike the d-cube, does not factor as a product of subgraphs, compression will not work. Something new is required. In the late 1970s, having successfully developed a notion of morphism for optimal flow problems (the subject of a projected second volume on global methods), I decided to take up the challenge. One of the things which had been learned in studying flowmorphisms was that symmetries induce flowmorphisms. A symmetry is a morphism, a trivial one in the sense of being a reduction since its range is identical to its domain.

As flowmorphisms though, symmetries also induce nontrivial flowmorphisms, the pushout with the identity. So, we were led to ask if there is a nontrivial notion of morphism for the EIP induced by symmetry? Abstracting from the earlier paper of Bernstein, Steiglitz and Hopcroft [**14**] produced the definition of stabilization in Section 1.2 of Chapter 1. That definition, and the futility of efforts to extend it to more general isomorphisms, established that reflections play a central role in the theory of isoperimetric problems. The graphs of the n-gon, d-simplex, d-cube and d-crosspolytope all have transitive symmetry groups generated by reflections which are stabilizing. What other diagram graphs have those properties?

H. S. M. Coxeter has written the classic monograph [**28**] about regular polytopes. One of the many fascinating facts there is a complete catalog of all the regular convex polytopes in all dimensions. This catalog was first established by another nineteenth century Swiss mathematician, L. Schläfli (actually a friend of J. Steiner, the inventor of symmetrization; see Chapter 2, Section 2.5 or the appendix). Coxeter's definition of a regular convex polytope [**28**, pp. 126–8] is inductive and technical but Schläfli's list and their well-known representations make them manifest. In all dimensions there are *the standard three*; the simplex (which is self-dual), the cube and the dual of the cube (crosspolytope). For $d > 4$ those are the only regular solids, but for $2 \leq d \leq 4$ there are some additional ones, which are called *exceptional*. They are listed in the table below.

d	Exceptional regular convex polytopes in $\mathbb{R}^d$
2	n-gons, $n \geq 5$
3	Dodecahedron Icosahadron
4	24-cell 120-cell 600-cell

The graph of the n-gon is $\mathbb{Z}_n$. The dodecahedron has 12 pentagonal faces, 30 edges and 20 vertices. The icosahedron, dual to the dodecahedron, has 20 triangular faces, 30 edges and 12 vertices and is familiar to mathematicians as the logo of the American Mathematical Association. We will describe the four-dimensional exceptional regular convex polytopes later. Anyway, the polytopes on Schläfli's list have the kind of graphs we are looking for. Their symmetry groups are transitive (on faces of all dimensions), generated by reflections (so they have lots of reflective symmetries) and all of those reflective symmetries are stabilizing (since convex polytopes have no crossing edges, see Theorem 3.1).

5.1.1 The dodecahedron and icosahedron

We have already calculated stability orders for $\mathbb{Z}_n$, Q_d and $Q_d^* = \maltese_d$. One may also do it for K_{d+1}, the graph of the d-simplex. Now consider the dodecahedron (with the aid of a model, if possible). It has 15 reflective symmetries whose fixed planes are the perpendicular bisectors of the 15 antipodal pairs of edges. Calculating the Hasse diagram of the stability order generated by all 15 reflections appears to be a daunting task, if one is doing it by hand and following the procedure of Section 2.4. However, in the calculations for $\mathbb{Z}_n$, etc. we learned that we could shorten the process by choosing wisely among the subsets of reflections with which to generate a stability order. We will shortly show how to choose the right subset in general, but for the purposes of this calculation we just indicate them (let's call them *basic* reflections) on the diagram in Fig. 5.1. The diagram is a projection of the graph of the dodecahedron onto the plane of one of its pentagonal faces. That face is in the center of the diagram. The projection is from a point on the line perpendicular to the pentagon and through its center. That line goes through the center of the dodecahedron and then passes through the center of the antipodal face. The point of projection is on that line, just after it passes through the center of the antipodal pentagon. That pentagon is thus projected onto a regular pentagon which is much enlarged and surrounds the whole diagram. The other 10 faces are also projected onto pentagons but their sides are lengthened or shortened by varying factors so that the images are not regular. The projections of the fixed planes for our basic reflections are indicated by dotted lines, labelled $\mathcal{R}_1$, $\mathcal{R}_2$ and $\mathcal{R}_3$. The fixed planes for $\mathcal{R}_1$ and $\mathcal{R}_2$ are perpendicular to the plane of projection so they project into straight lines but $\mathcal{R}_3$'s fixed plane is transverse so its dotted lines are the projection of the intersections of the plane with the faces of the dodecahedron. They form a closed curve, a nonregular hexagon, but two of its sides coincide with edges which hide the dotted lines. The dotted lines for $\mathcal{R}_1$ and $\mathcal{R}_2$ are actually fixed lines for reflective symmetries of the projected diagram so the actions of $\mathcal{R}_1$ and $\mathcal{R}_2$ on the vertices are easily determined.

Even though the dotted lines for $\mathcal{R}_3$ are not straight, one can still determine from them the action of $\mathcal{R}_3$ on the vertices and that is all that we need in order to calculate the stability order. The Fricke–Klein point, p, is in the small triangle formed by the projections of the three planes. Actually, the basic reflections were chosen to be closest, in that sense, to p.

So we begin by computing the stability order of the dodecahedron wrt this relatively small subset of reflective symmetries. There is only one vertex on the p-side of all three planes. We label it a since it must be a minimal element of the stability order. We proceed from there, scanning $\mathcal{R}_1(a)$, $\mathcal{R}_2(a)$ and $\mathcal{R}_3(a)$,

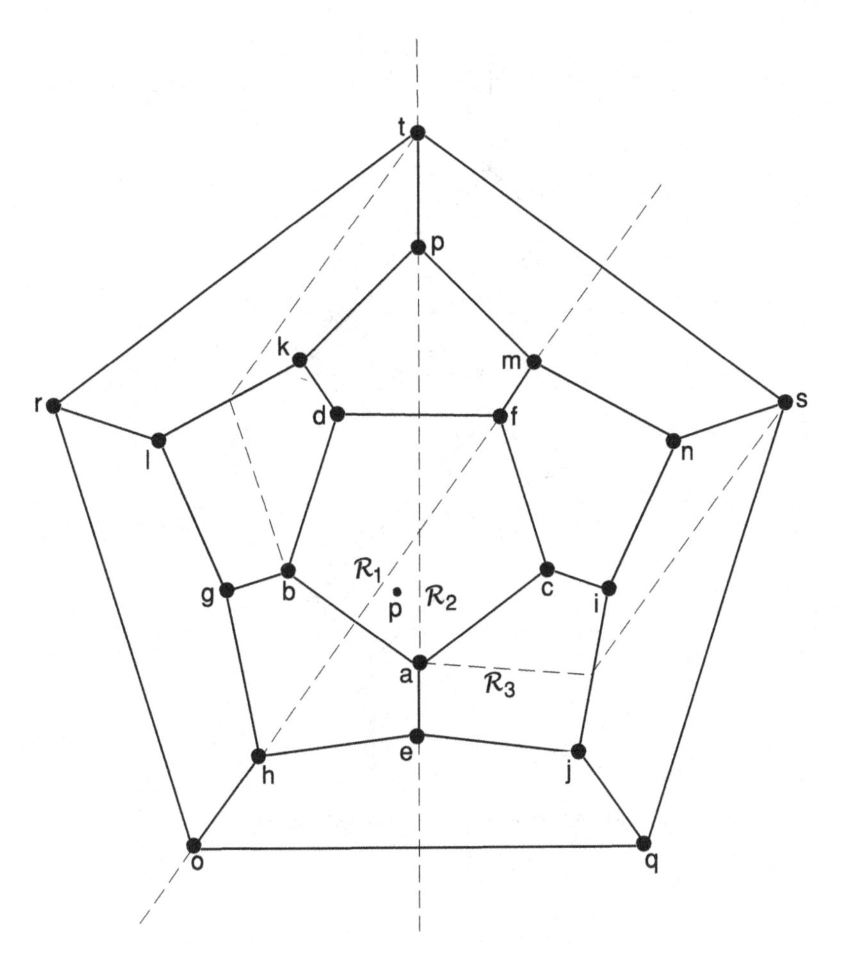

Fig. 5.1 Planar projection of the dodecahedron.

in that order. $\mathcal{R}_1(a) \neq a$ so we label it b. $\mathcal{R}_2(a) = a = \mathcal{R}_3(a)$ so they give nothing new. $\mathcal{R}_1(b) = a$, which has already been labeled, but $\mathcal{R}_2(b)$ has not been labeled so we label it c. $\mathcal{R}_3(b) = b$ so it gives nothing new. Proceeding in this manner we produce the diagram (solid lines, labeled with reflections) of Fig. 5.2. Since this diagram includes all vertices, it must be the Hasse diagram of $\mathcal{S}(V_{20}; \mathcal{R}_1, \mathcal{R}_2, \mathcal{R}_3; p)$. Even by hand we should be able to calculate the derived network of this stability order, but remember that we have only used 3 out of 15 reflections and there seems to be obvious places in the diagram where an edge is missing (the dotted lines from f to i and l to o). Looking at those pairs in Fig. 5.1, we see that there is a reflection (nonbasic) which takes f to

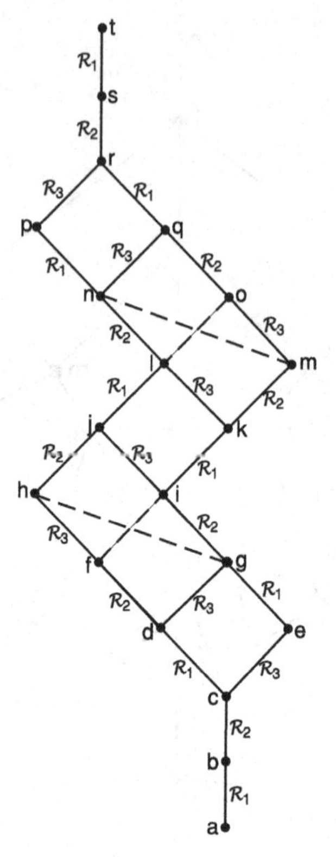

Fig. 5.2 Stability order of the dodecahedron.

i and l to o. Checking how that reflection maps the rest of the vertices, we see that a, c, r and t are fixed. Of the mirror image pairs, (b, e), (d, j), (k, q) and (p, s) are already comparable in the diagram but (g, h) and (m, n) are not, so we add them in too (the other pair of dotted lines). One can check the remaining incomparable pairs and show that none of them are images under any reflective symmetry of the dodecahedron. Thus we have computed the stability order of the dodecahedron wrt all of its reflective symmetries.

From that stability order we calculate the derived network (Fig. 5.3). In Fig. 5.3 the minimum Θ-value for each cardinality is in boldface. Note that the icosahedron has nested solutions for the EIP and they are just as expected. There is though, one stable solution set, $\{a, b, c, e\}$, which is not contained in any solution 5-set.

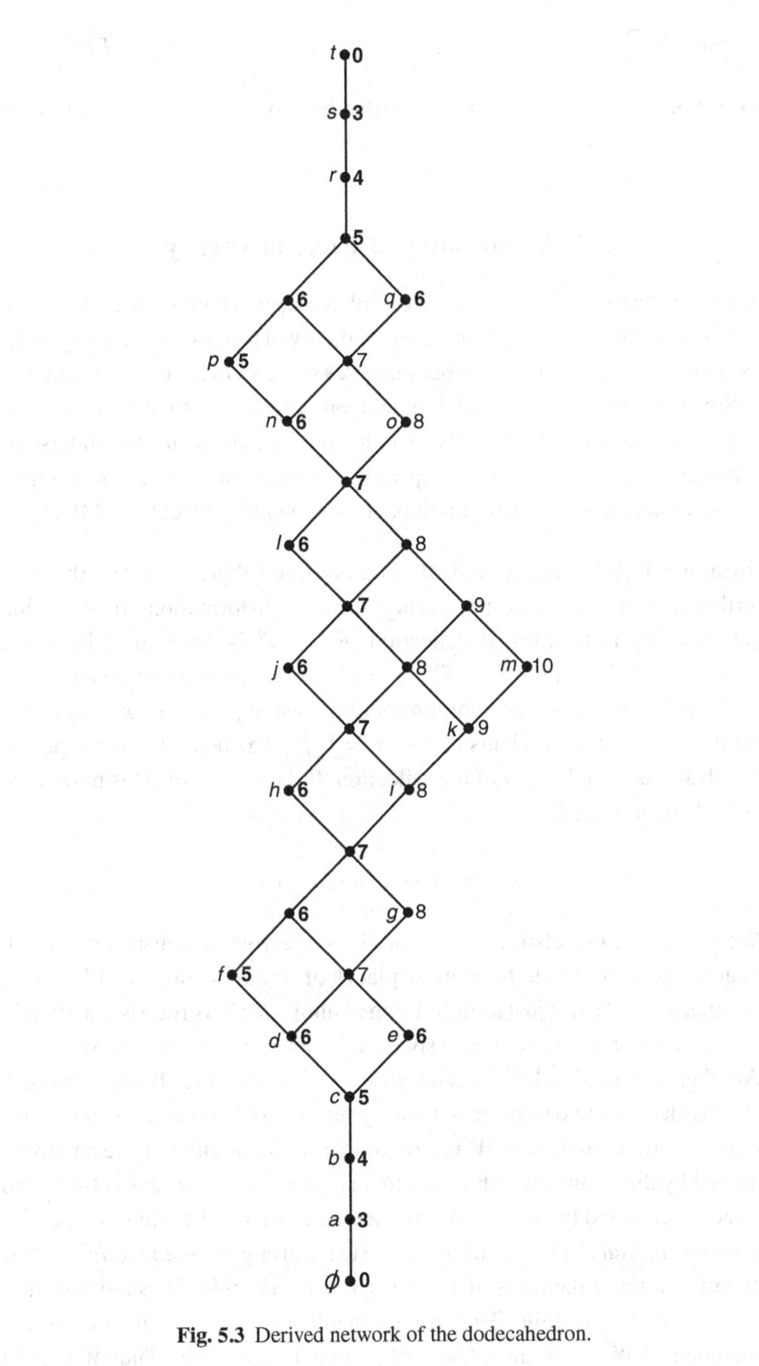

Fig. 5.3 Derived network of the dodecahedron.

Exercise 5.1 *Solve the VIP for the dodecaheron as we solved the EIP.*

Exercise 5.2 *Solve the EIP for the icosahedron as we have for the dodecahedron.*

5.2 A summary of Coxeter theory

Coxeter [**28**] beginning just as we have, with the regular solids and their symmetry groups, laid the foundation for a general theory of groups generated by reflections. This is a wonderful tool for our purposes so we shall devote some attention to it. Since we are using it as a tool though, our intention is to present just those results which are relevant and only include proofs as necessary for understanding. Fortunately, there are many expositions where one can get the complete theory with details and different points of view on the subject (cf. [**23**] or [**10**])

Definition 5.1 A *Coxeter group, W,* is a discrete subgroup of O_d, the set of all orthogonal (i.e. distance-preserving) affine transformations of $\mathbb{R}^d$, which is generated by reflections. A reflection, $\mathcal{R}$, *of* $\mathbb{R}^d$ is determined by a *fixed hyperplane,* $H(\mathcal{R}) = \left\{x \in \mathbb{R}^d : \mathcal{R}(x) = x\right\}$. A hyperplane (affine subspace of dimension $d-1$) is also the solution set of a linear equation $e \cdot x = c$, $e \in \mathbb{R}^d$ being the vector of coefficients and $e \cdot x = \sum_{i=1}^{d} e_i x_i$ being the inner product of e with x. e is called a *root* of the reflection. If the root, e, of $\mathcal{R}$ is normalized $(e \cdot e = 1)$ then $\forall x \in \mathbb{R}^d$

$$\mathcal{R}(x) = x - 2(e \cdot x - c)\, e.$$

We will generally also assume that W is *effective*, which means that the orthogonal vectors of (the fixed hyperplanes of) the reflections in W span the whole space, $\mathbb{R}^d$. This is just a simplifying assumption; if a group is not effective then its restriction to the subspace spanned by its roots will be effective.

Another way to simplify Coxeter groups is by factoring: If the roots of W can be divided into two or more nonempty subsets so that members of different subsets are orthogonal, then W is isomorphic to the product of the subgroups generated by those subsets (which are themselves Coxeter groups acting on the subspaces generated by their respective sets of roots and the whole space is the *rectangular product* of those subspaces). Such a group is called *reducible*. If no such partitioning of the roots of W exists, it is *irreducible*. We shall assume in this chapter that W is finite. This implies that there is a point which is fixed by all members of W and we may take that point to be the origin. Thus W consists of linear transformations.

Examples are:

(1) The dihedral group D_2^n, the symmetry group of $\mathbb{Z}_n$ (actually the dihedron, a disk bounded by a regular n-gon), is a Coxeter group. $D_2^2 = D_2^1 \times D_2^1$ but D_2^n is irreducible if $n > 2$.
(2) The cuboctahedral group, the symmetry group of $\mathbb{Q}_d$ and $\mathbb{Q}_d^*$, is an irreducible Coxeter group for every d.

5.2.1 Basic facts

5.2.1.1 Chambers

Removing the fixed hyperplanes of all reflections in W, the rest of $\mathbb{R}^d$ is partitioned into connected components called *chambers.* Coxeter showed that each chamber is a simplex. Since, under the assumption of finiteness for W, the fixed hyperplanes are linear subspaces, the simplex has one face at infinity and would more properly be called a simplicial cone. The chambers of reducible Coxeter groups are rectangular products of the simplicial cones of its irreducible factors. Let $p \in \mathbb{R}^d$ be any point not fixed by any reflection of W and call it the Fricke–Klein point. The unique chamber containing the Fricke–Klein point is called the *fundamental chamber*, C_0. A basic fact is that W acts transitively on chambers. In fact for any chamber C there is a unique $g \in W$ such that $g(C_0) = C$ (see [**10**], Section 4.2.1). This determines a one-to-one correspondence between chambers and elements of W.

Exercise 5.3 *Show that any reflection in a Coxeter group is conjugate to a unique basic reflection.*

In general, a Coxeter group may contain stabilizing and nonstabilizing reflections but the foregoing exercise shows that if W is generated by stabilizing reflections, then all reflections in W are stabilizing.

5.2.1.2 Generators and relations

Coxeter showed that the reflections $\mathcal{R}_1, \mathcal{R}_2, ..., \mathcal{R}_d$, whose fixed hyperplanes bound C_0, are a minimal generating set for W (called the *basis*) and they satisfy the relations

(1) $(\mathcal{R}_i)^2 = \mathcal{I}$, the identity, for all i, and
(2) $(\mathcal{R}_i\mathcal{R}_j)^{m_{ij}} = \mathcal{I}$, m_{ij} being the order of $\mathcal{R}_i\mathcal{R}_j$, for $i \neq j$. The composition of two distinct reflections is, geometrically, a rotation about $H(\mathcal{R}_i) \cap H(\mathcal{R}_j)$ of angle $2\cos^{-1}(e_i \cdot e_j)$. Thus $m_{ij} = \frac{\pi}{|\cos^{-1}(e_i \cdot e_j)|}$.

Witt showed that all other relations are a consequence of these (see [**28**] pp. 80–81 for details) which led to the abstract definition and subsequent generalization of Coxeter groups (see Chapter 1 of [**54**], a condensation of Chapters IV, V and VI of [**23**]).

5.2.1.3 Classification

Coxeter noted that a Coxeter group was determined by the shape of its fundamental chamber which could be characterized by a simple graph. The vertices of this graph represent the (finite) faces of the fundamental chamber with an edge between two vertices if their roots are not orthogonal. The angle between them must be $\frac{\pi}{m}$ radians, $m = m_{ij} > 2$, so we label the edge m (by convention, if $m = 3$ we leave the label off). Note that this graph is connected if and only if W is irreducible.

From the relation between these graphs and a quadratic form defined by the group, Coxeter was able to show that there were just four possible infinite families of finite irreducible Coxeter groups and six possible exceptional ones. Coxeter's catalog of the finite irreducible Coxeter groups is closely related to Schläfli's catalog of regular convex polytopes (Section 5.1) which is an extension of Euclid's catalog for the two- and three-dimensional regular solids. The existence for all of these possibilities was demonstrated by construction. Diagrams of their Coxeter graphs and the order of the group are given in Table 5.1 (taken from Table IV on p. 297 of [**28**] but with the slightly different notation of Cartan which is now standard; see p. 32 of [**56**]). A_d is the symmetry group of the d-simplex which is isomorphic, as a group, to the symmetric group, S_{d+1}. B_d is the cuboctohedral group. C_d is closely related to B_d, being the symmetry group of the even vertices of $\mathbb{Q}_d$. G_3 is the common symmetry group of the dodecahedron and its dual, the icosahedron.

5.2.1.4 Length

If W is a Coxeter group with basic reflections $\mathcal{R}_1, \mathcal{R}_2, \ldots, \mathcal{R}_d$, then any $g \in W$ will have a representation

$$g = \mathcal{R}_{i_1} \circ \mathcal{R}_{i_2} \circ \ldots \circ \mathcal{R}_{i_\ell},$$

and generally g will have many such representations. The smallest integer, $\ell(g)$, such that g may be written as such a composition of $\ell = \ell(g)$ basic reflections will be called the *length of* g. Such a minimal representation of g (with $\ell(g)$ basic reflections) is called a *reduced expression* for g.

Table 5.1. *finite irreducible coxeter groups*

Symbol	Coxeter graph	Order
A_d		$(d+1)!$
B_d	 4	$2^d d!$
C_d	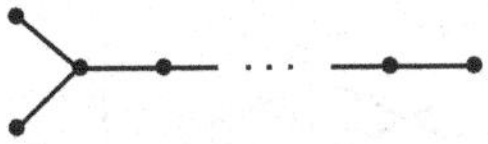	$2^{d-1}d!$
E_6	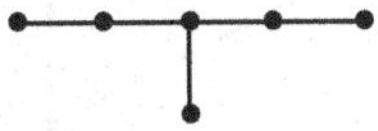	$72 \cdot 6! = 51\,840$
E_7	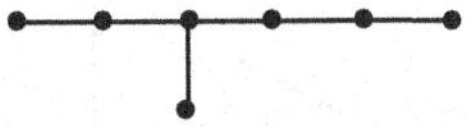	$8 \cdot 9! = 2903\,040$
E_8		$192 \cdot 10! = 696\,729\,600$
F_4	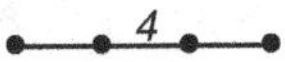 4	1152
G_3	5	120
G_4	 5	$120^2 = 14400$
$I_2(n)$	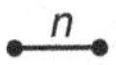 n	$2n$

5.2.1.5 Bruhat order

If W is a (finite) Coxeter group with basic reflections $\mathcal{R}_1, \mathcal{R}_2, ..., \mathcal{R}_d$, and $g, h \in W$, we say *h is less than g* (notation: $h < g$) if there is a conjugate, $\mathcal{R}$, of some basic reflection (i.e. $\mathcal{R}$ is a reflection but not necessarily a basic reflection) such that $h = \mathcal{R} \circ g$ and $\ell(h) < \ell(g)$. The *Bruhat order of W* is the transitive closure of $<$.

5.2.1.6 Example

The Hasse diagram of the Bruhat order of the dihedral group, $I_2(4)$, is shown in Fig. 5.4. The solid edges represent basic reflections and the dotted edges nonbasic reflections.

Exercise 5.4 *Which reflection gives the covering relation $\mathcal{R}_1 \to \mathcal{R}_2 \circ \mathcal{R}_1$? Which gives $\mathcal{R}_1 \to \mathcal{R}_1 \circ \mathcal{R}_2$?*

The key to analyzing Bruhat order is the following theorem, called the Exchange Condition which relates Bruhat order to the reduced decompositions of

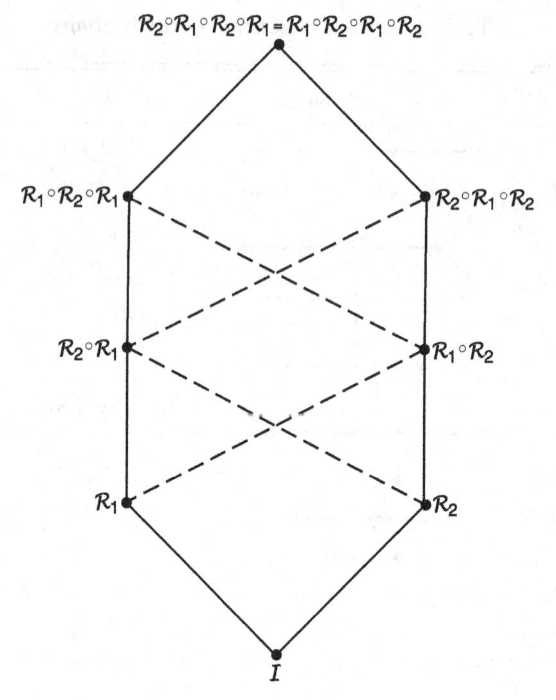

Fig. 5.4 Hasse diagram of the Bruhat order of $I_2(4)$.

group elements. The Exchange Condition was discovered by H. Matsumoto for basic reflections and then extended to all reflections by D. N. Verma (see [**54**], Proposition 6.1 of Chapter 1 for a proof).

Theorem 5.1 *Let* $g = \mathcal{R}_{i_1} \circ \mathcal{R}_{i_2} \circ \ldots \circ \mathcal{R}_{i_n}$ *be a decomposition of* g *and* $\mathcal{R}$ *any reflection in* W. *If* $\ell(\mathcal{R} \circ g) \leq \ell(g)$, *then there is a* k, $1 \leq k \leq n$, *such that*

$$\mathcal{R} \circ \mathcal{R}_{i_1} \circ \ldots \circ \mathcal{R}_{i_k} = \mathcal{R}_{i_1} \circ \ldots \circ \mathcal{R}_{i_{k-1}}.$$

Corollary 5.1 *Length is a rank function on the Bruhat order of any Coxeter group.*

5.2.1.7 Parabolic subgroups

Let A be any subset of $\Omega = \{\mathcal{R}_1, \mathcal{R}_2, \ldots, \mathcal{R}_d\}$, the set of basic reflections. The subgroup of W generated by A is a Coxeter group (wrt the space spanned by its roots), W_A. Such subgroups are called *parabolic subgroups* of W. The Coxeter graph of W_A is the subgraph of the Coxeter graph of W induced by A.

In addition,

$$W_A \cap W_B = W_{A \cap B}$$

and the subgroup generated by $W_A \cup W_B$ is $W_{A \cup B}$. Thus the parabolic subgroups of a finite Coxeter group, W of rank d, form a Boolean lattice of size 2^d with $W_\emptyset = \{\mathcal{I}\}$ and $W_\Omega = W$. In fact there is an isomorphism, F, between this Boolean lattice and the Boolean lattice of finite faces of the fundamental chamber:

$$F(A) = C_0 \cap \bigcap_{\mathcal{R}_i \in A} H(\mathcal{R}_i).$$

5.3 The structure of stability orders

Now, suppose that we have a diagram graph, G, which has some stabilizing symmetry. We may as well utilize all of the stabilizing symmetries of G since that will make the derived network smaller and the subsequent calculations easier. These stabilizing symmetries, being reflections, will generate a group, W. If W is not effective then we can isomorphically project G onto the hyperplane generated by its orthogonal vectors so that the elements of W are symmetries of the image of G. We may assume then that W is a Coxeter group.

W might not be the full symmetry group of G and there might even be other reflective symmetries of G which are not stabilizing, but since every reflection in W is conjugate to one of its generators, all reflections in W must be stabilizing. We denote the stability order of G wrt all reflections of the Coxeter group, W, and the Fricke–Klein point, p, by $\mathcal{S}(G; W; p)$.

Theorem 5.2 *The stability order $\mathcal{S}(G; W; p)$ is independent of the point p; i.e.*

(1) *if p and q are in the same chamber, then $\mathcal{S}(G; W; p) = \mathcal{S}(G; W; q)$, and*
(2) *if p and q are in different chambers, then $\mathcal{S}(G; W; p) \simeq \mathcal{S}(G; W; q)$, i.e. they are isomorphic.*

Proof Part (1) follows from the observation that $P^{(1)}(G; W; p) = P^{(1)}(G; W; q)$, i.e. for any $v \in V_G$ and $\mathcal{R} \in W$, $\|v - p\| < \|\mathcal{R}(v) - p\|$ if and only if $\|v - q\| < \|\mathcal{R}(v) - q\|$. For part (2), assume $p \in C_0$ and $q \in C_0'$. Then, by Coxeter there is a unique $g \in W$ such that $g(C_0) = C_0'$. Since g is a linear automorphism of G, its restriction to V_G is one-to-one and onto.

Claim 5.1 *As a map from $\mathcal{S}(G; W; p)$ to $\mathcal{S}(G; W; g(p))$, g is order-preserving and thus an isomorphism.*

Proof Suppose that $(v, \mathcal{R}(v)) \in P^{(1)}(G; W; p)$, i.e. $\mathcal{R}$ is a stabilizing reflection of G and $\|v - p\| < \|\mathcal{R}(v) - p\|$. By elementary linear algebra, $\mathcal{R}' = g \circ \mathcal{R} \circ g^{-1}$ is also a reflection and is in W so must be stabilizing. Therefore

$$\begin{aligned}
\|g(v) - g(p)\| &= \|g(v - p)\|, \ g \text{ being linear,} \\
&= \|v - p\|, \ g \text{ being distance-preserving,} \\
&< \|\mathcal{R}(v) - p\|, \ \text{by hypothesis,} \\
&= \|g(\mathcal{R}(v) - p)\| = \|g \circ \mathcal{R}(v) - g(p)\| \\
&= \left\|g \circ \mathcal{R} \circ g^{-1}(g(v)) - g(p)\right\| \\
&= \left\|\mathcal{R}'(g(v)) - g(p)\right\|.
\end{aligned}$$

Therefore $\left(g(v), \mathcal{R}'(g(v))\right) \in P^{(1)}(G; W; g(p))$ so we have proven the claim. □

Since $g(p) \in C_0'$, part (2) of the theorem follows from part (1). □

5.3.1 The components of stability orders

5.3.1.1 Definition

If $\mathcal{P}$ is a poset then $x, y \in \mathcal{P}$ are called *related* if $x \leq_{\mathcal{P}} y$ or $y \leq_{\mathcal{P}} x$. $x, y \in \mathcal{P}$ are called *connected* if there exists a sequence, $z_0, z_1, \ldots, z_k$, such that $x = z_0$, $y = z_k$ and each consecutive pair, z_{i-1}, z_i is related. In other words, connectivity is the smallest equivalence relation containing relativity. $\mathcal{P}$ itself is called *connected* if every pair of elements in it is connected. The maximal connected subsets of a poset are called its *components*.

Theorem 5.3 *Every poset has a unique representation as the direct sum (disjoint union) of its components.*

Proof This is a standard fact from order theory (see [**21**]). □

Theorem 5.4 *A component of the stability order $\mathcal{S}(G; W; p)$ has exactly one vertex in each chamber.*

Proof Any sequence of reflections, $\mathcal{R}_{i_1}, \mathcal{R}_{i_2}, \ldots, \mathcal{R}_{i_k}$, which relate vertex v to vertex w determine a symmetry $g = \mathcal{R}_{i_k} \circ \ldots \circ \mathcal{R}_{i_2} \circ \mathcal{R}_{i_1}$ such that $g(v) = w$. Conversely, given $g \in W$ there is a representation of g as a composition of reflections. The theorem then follows from the remark that if C and C' are

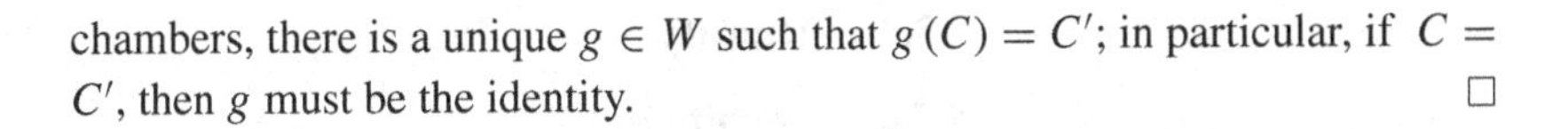

chambers, there is a unique $g \in W$ such that $g(C) = C'$; in particular, if $C = C'$, then g must be the identity. □

Theorem 5.5 *Every component of $\mathcal{S}(G; W; p)$ has a unique minimal element.*

Proof By the previous theorem the given component has a unique vertex in the fundamental chamber; call it v_0. Since v_0 and p are on the same side of the fixed hyperplane of any reflection $\mathcal{R}$, if $v_0 \neq \mathcal{R}(v_0)$, then $v_0 < \mathcal{R}(v_0)$. Therefore v_0 is minimal. Suppose w is any other vertex in the given component of $\mathcal{S}$. Since w cannot be in the fundamental chamber, there must be a fundamental reflection, $\mathcal{R}_i$, such that p is on one side of its fixed hyperplane and w is on the other. But then $|\mathcal{R}(w) - p| < |w - p|$. Therefore $\mathcal{R}(w) < w$ and w is not minimal. □

According to Coxeter theory, the fundamental chamber of a finite Coxeter group is a simplex with one face at infinity. Assuming that $\mathcal{S}(G; W; p)$ is connected, its unique minimal element will determine a face of the fundamental chamber, the smallest face which contains it. There are $2^{d+1} - 2^d = 2^d$ faces on the fundamental chamber.

Theorem 5.6 *If $\mathcal{S}(G; W; p)$ and $\mathcal{S}(H; W; p)$ are connected and their minimal elements determine the same face of the fundamental chamber, then $\mathcal{S}(G; W; p) \simeq \mathcal{S}(H; W; p)$.*

Proof The smallest face of a simplex containing a particular point must contain that point in its interior. The intersection of a chamber and a fixed hyperplane is a face of the chamber. Thus the intersection of the common smallest face with any fixed hyperplane will be the face itself or some subface of it. In either case the reflection will map both minimal elements in the same way and similarly with their images. □

In [**54**], Chapter 1, Section 5, these isomorphic stability orders are identified with quotients of the Bruhat order on W by a parabolic subgroup, W_A, the group of members of W which fix the common face of the fundamental chamber determined by the minimal elements of G and H. Hiller [**54**] calls them *Bruhat posets* so we may sum all this up with

Theorem 5.7 *Every stability order $\mathcal{S}(G; W; p)$ is isomorphic to the disjoint union of Bruhat posets.*

5.3.2 Duality

5.3.2.1 Definition

If $\mathcal{P} = (P, \leq)$ is a poset, then the poset $\mathcal{P}^* = (P, \geq)$ is called the *dual* of $\mathcal{P}$. If $\mathcal{P} \simeq \mathcal{P}^*$, then $\mathcal{P}$ is called *self-dual.*

Theorem 5.8 *$\mathcal{S}(G; W; p)$ is self-dual.*

Proof $\mathcal{S}^*(G; W; p) = \mathcal{S}(G; W; -p)$ since p and $-p$ lie on opposite sides of the fixed hyperplane of any reflection. The theorem follows by Theorem 5.2. □

5.4 Calculating stability orders

5.4.1 The algorithm

The definition of stability order, given in Section 3.2.7.1, is not really good from the standpoint of computation. For a graph G on n vertices whose stabilizing reflections generate a Coxeter group, W, calculating the Hasse diagram of $\mathcal{S}(G; W; p)$ from its definition could take $O\left(n^4\right)$ steps. In the examples following the definition, contrary to the remark at the beginning of Section 5.3, we did not really use all of the stabilizing symmetries, but a judiciously chosen subset. Now that we know a little Coxeter theory, we can see that those helpful subsets of reflections were Coxeter bases. If we restrict ourselves to a Coxeter basis, $\mathcal{R}_1, \mathcal{R}_2, ..., \mathcal{R}_d$, of W, and a vertex v_0 in the fundamental chamber, then the calculation of its component of the stability order (called a *weak* Bruhat *order*) is greatly simplified. Because of linear independence, the Hasse diagram of $\mathcal{S}(G; \mathcal{R}_1, \mathcal{R}_2, ..., \mathcal{R}_d; p)$ is just $P^{(1)}$ which takes only $O(n)$ steps. For $\mathbb{Z}_n$, $\mathcal{Q}_d$ and $\mathcal{Q}_d^*$, we were able to see in looking back that $\mathcal{S}(G; \mathcal{R}_1, \mathcal{R}_2, ..., \mathcal{R}_d; p) = \mathcal{S}(G; W; p)$, the *strong* Bruhat order, so our shortcut cost us nothing. But in general $\mathcal{S}(G; \mathcal{R}_1, \mathcal{R}_2, ..., \mathcal{R}_k; p) \subseteq \mathcal{S}(G; W; p)$ and the inclusion, as we saw with the dodecahedron, may be strict.

Corollary 5.2 *(of Theorem 5.1) Length is a rank function for the weak, as well as the strong, Bruhat order.*

Thus the only difference between the Hasse diagram of $\mathcal{S}(G; W; p)$ and that of $\mathcal{S}(G; \mathcal{R}_1, \mathcal{R}_2, ..., \mathcal{R}_k; p)$ (which is relatively easy to compute) is a few extra edges between elements of consecutive ranks. Having the Hasse diagram of $\mathcal{S}(G; \mathcal{R}_1, \mathcal{R}_2, ..., \mathcal{R}_k; p)$, these can be easily added to obtain the Hasse diagram of $\mathcal{S}(G; W; p)$. Altogether, the observations of this section and the preceding one give a simple two-step process for constructing the Hasse diagram of $\mathcal{S}(G; W; p)$. Let $\mathcal{S}_\ell = \{v \in V : \ell(v) = \ell\}$ and assume that $\mathcal{S}(G; W; p)$ is connected. If not, just repeat for each component.

(1) Partition V into ranks, $\mathcal{S}_\ell$, using the weak stability order:
   (a) Begin with the unique vertex, v_0, in the fundamental chamber. $\mathcal{S}_0 = \{v_0\}$.
   (b) Extend from $\mathcal{S}_\ell$ to $\mathcal{S}_{\ell+1}$ by applying each of the basic reflections to each member of $\mathcal{S}_\ell$. The result will either be in $\mathcal{S}_{\ell-1}$ or $\mathcal{S}_{\ell+1}$, so we need only eliminate those we know to be in $\mathcal{S}_{\ell-1}$ to get $\mathcal{S}_{\ell+1}$.
   (c) When $\mathcal{S}_{\ell+1} = \emptyset$ we are done.
(2) Generate the Hasse diagram of $\mathcal{S}(G; W; p)$. Examine all pairs (v, w) with $v \in \mathcal{S}_\ell$ and $w \in \mathcal{S}_{\ell+1}$, to see if $v <_S w$, i.e. if there exists a reflection, $\mathcal{R} \in W$ such that $\mathcal{R}(v) = w$. Those for which it does, make up the Hasse diagram of $\mathcal{S}(G; W; p)$.

5.4.2 The deBruijn graph revisited

In Section 1.3.2 of Chapter 1 we considered the wirelength problem on the deBruijn graph of order 4, deducing its solution from the assumption that we had a solution to the EIP on the graph. Fig. 5.5 shows the diagram of the deBruijn graph of order 4 with the optimal numbering on the vertices (see Fig. 1.5 in Chapter 1). Also marked by dotted lines are the fixed lines of two reflective symmetries, $\mathcal{R}_1$ and $\mathcal{R}_2$.

Given the representation of the vertices of DB_4 as 4-tuples of 0s and 1s (as shown in Fig. 5.4), $\mathcal{R}_1$ is induced by interchanging 0 and 1. $\mathcal{R}_2$ is induced by reversing the order of digits in the 4-tuples. These two reflections generate the reducible Coxeter group $D_2^2 \simeq D_2^1 \times D_2^1$. Choosing the lower left quadrant as the fundamental chamber, it contains the vertices labeled 1, 2, 4, 6, 7, 10 so the stability order has six components with these as their minima. That stability order is shown in Fig. 5.6, and the resulting derived network, represented as a product and not including weights, is in Fig. 5.7.

Since it has $3^4 \times 6^2 = 2916$ vertices (each representing a stable set) a diagram of the derived network itself is not feasible. However it is a straightforward matter to have a computer generate them and evaluate $|\Theta(S)|$ for each one. The solutions of the EIP were found to be

k	0	1	2	3	4	5	6	7	8	9	10	11	12	13	14	15	16
$\min_{S \subseteq V, \lvert S\rvert = k} \lvert\Theta(S)\rvert$	0	2	4	4	6	6	6	6	6	6	6	6	6	4	4	2	0

This verifies the solution of the wirelength problem for DB_4 claimed in Section 1.3.2. Brute force would have required the examination of

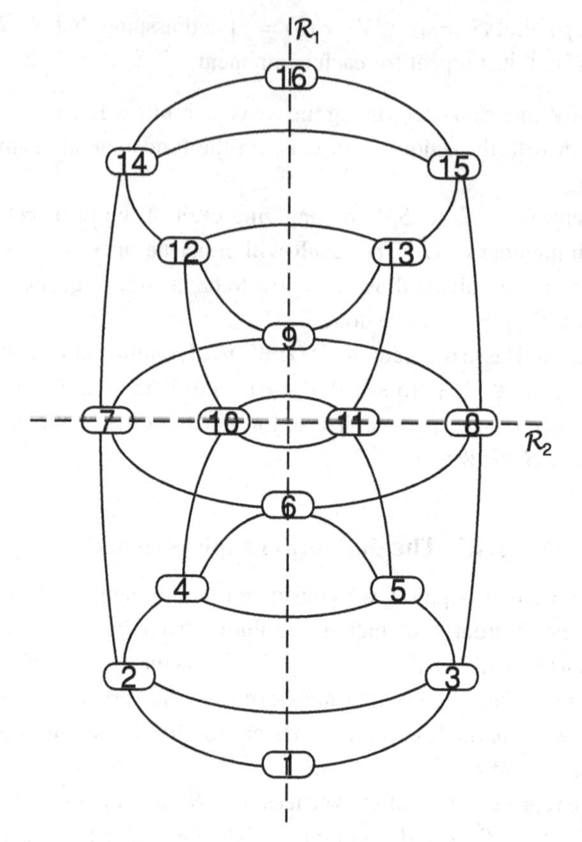

Fig. 5.5 DB_4 with stabilizing symmetries.

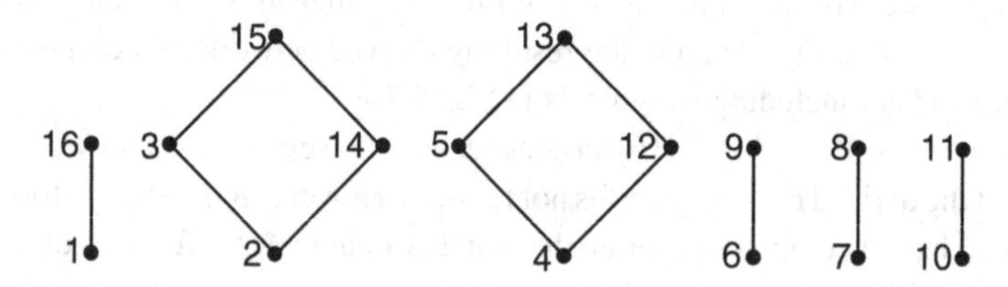

Fig. 5.6 Stability order of DB_4.

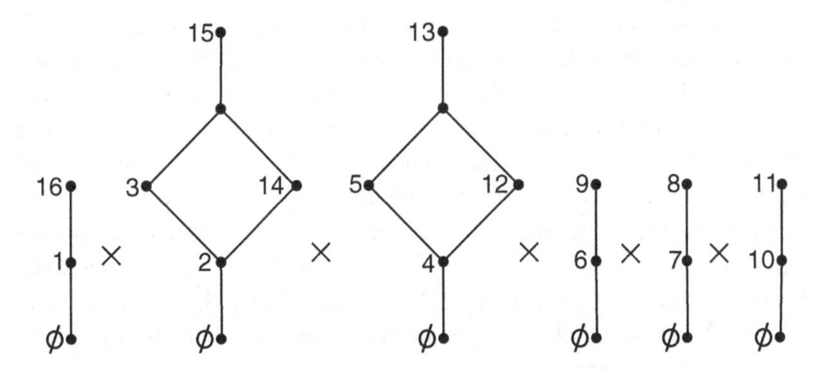

Fig. 5.7 The Hasse diagram of the derived network of DB_4.

$2^{16} = 65\,536$ subsets so a significant saving, a more than 20-fold reduction, was achieved.

5.5 Into the fourth dimension

5.5.1 The 24-vertex (24-cell)

Now we apply our theory of stabilization to isoperimetric problems on the graphs of the four-dimensional exceptional regular solids. According to Schläfli (see Section 5.1) there are three of them; we begin with the smallest. Coxeter [**28**] calls it the 24-cell since it is made up of 24 octahedra which fit together in $\mathbb{R}^4$, joining at their triangular faces to make a regular solid. Its Schläfli symbol (see [**28**]) is $\{3, 4, 3\}$ which, being palindromic indicates that the 24-cell is self-dual and has 24 vertices also. The number of vertices is more significant for combinatorial purposes, so we shall refer to it, and the other exceptional regular solids, by the number of its vertices. Coxeter [**28**] describes constructions for all of the exceptional regular solids in $\mathbb{R}^4$ (starting from Q_4), which lead to coordinates for their vertices (assuming the standard coordinates, $\{1, -1\}^4$, for Q_4). Cesaro's construction for $\{3, 4, 3\}$ (Section 8.2 of [**28**]) gives its 24 vertices as the midpoints of the edges of $Q_4^* = \maltese_4$. Recall that $V_{\maltese_4} = \{\pm\delta_i : 1 \le i \le 4\}$, so $V_{24} = \left\{\pm\delta_i \pm \delta_j : 1 \le i < j \le 4\right\}$ (we have left off the irrelevant factor of $\frac{1}{2}$). An equivalent description of V_{24} is as the set of all permutations of $(0, 0, \pm1, \pm1)$. The edges of this 24-vertex are line segments connecting pairs of vertices at minimum distance which is easily seen to be $\sqrt{2}$. Thus the neighbors of $(0, 0, 1, 1)$ are $(0, \pm1, 1, 0)$, $(\pm1, 0, 1, 0)$, $(0, \pm1, 0, 1)$, $(\pm1, 0, 0, 1)$. Each vertex has eight neighbors so there are $(24 \times 8)/2 = 96$ edges altogether.

One way to calculate the stability order of V_{24}, and the way that it was done originally [**48**], is to work with a diagram (model) of a projection of V_{24} into $\mathbb{R}^3$, just as we did with a diagram of a projection of the dodecahedron into $\mathbb{R}^2$. One might even be able to do it with the diagrams of projections of V_{24} into $\mathbb{R}^2$ given in Coxeter, but we were fortunate in finding a ready-made three-dimensional model which was perfect for the purpose, and the whole calculation took no more time than it had for the dodecahedron. However, that approach will not work for the 120-vertex, much less the 600-vertex. Coxeter's book has some pictures of models of projections of V_{120} (Plate IV of [**28**]) and V_{600} (Plate V) into $\mathbb{R}^3$, but they look like balls of twine; very discouraging for anyone contemplating such a project.

It should now be apparent, however, that the whole process can be programmed on the computer. The computer has no geometric intuition and would not really "understand" a diagram, but it is able to deal very well with 4-tuples of real numbers representing vertices. The reflections can be represented by root vectors or matrices and the whole process becomes an exercise in linear algebra. In order to demonstrate how this works, we now run through it for the 24-vertex (for which a computer is not necessary and all calculations can be done by hand).

Having V_{24} and knowing which pairs of vertices are connected by edges, the next thing we need is the hyperplanes of reflective symmetry, i.e. their roots. One may determine these by brute force. Every reflective symmetry, $\mathcal{R}$, must take some vertex, x, to another, $y = \mathcal{R}(x) = x - 2(x \cdot e)e$, where e is the normalized root of $\mathcal{R}$. Therefore

$$e = \frac{x - \mathcal{R}(x)}{\|x - \mathcal{R}(x)\|}.$$

To find all the reflections then, we could just take the difference $x - y$ for all $x, y \in V$, normalize and see if the resulting reflection is a symmetry of V. For V_{24} this is a modest number of calculations, $\binom{24}{2} = 276$, not what one would want to do by hand but no problem for the computer. However, as is often the case in this business, there is an easier way and one which provides more insight. In Section 12.6 of [**28**] it is pointed out that V_{24} has 24 reflective symmetries. For 12 of them the root direction goes through a pair of antipodal vertices. Choosing the one whose first nonzero entry is 1, we have the roots (not normalized)

$$(0, 0, 1, 1),$$
$$(0, 0, 1, -1),$$
$$(0, 1, 0, 1),$$
$$(0, 1, 0, -1),$$
$$(0, 1, 1, 0),$$
$$(0, 1, -1, 0),$$
$$(1, 0, 0, 1),$$
$$(1, 0, 0, -1),$$
$$(1, 0, 1, 0),$$
$$(1, 0, -1, 0),$$
$$(1, 1, 0, 0),$$
$$(1, -1, 0, 0).$$

The other 12 have their roots through the centers of octahedral faces (vertices of the dual 24-vertex). Cesaro [28, Section 8.2] found those 24 octahedral cells by truncating Q_4^*. Eight of them come from the vertex figures of Q_4^*, whose centers are the vertices of Q_4^*. Again, these occur in antipodal pairs and we take the one whose nonzero entry is 1,

$$(0, 0, 0, 1),$$
$$(0, 0, 1, 0),$$
$$(0, 1, 0, 0),$$
$$(1, 0, 0, 0).$$

The other 16 come from the truncations of the tetrahedral cells of Q_4^* whose centers are the vertices of Q_4 and as before they occur in antipodal pairs from which we select the one with first entry 1,

$$(1, 1, 1, 1),$$
$$(1, -1, 1, 1),$$
$$(1, 1, -1, 1),$$
$$(1, -1, -1, 1),$$

$$
\begin{gathered}
(1,1,1,-1)\,,\\
(1,-1,1,-1)\,,\\
(1,1,-1,-1)\,,\\
(1,-1,-1,-1)\,,
\end{gathered}
$$

completing the list. Note that those of the first group have length $\sqrt{2}$, the second, 1, and the third, 2.

The next step is to select a Fricke–Klein point, p. The defining property of p is that it does not lie on the fixed hyperplane for any reflection, i.e. for all roots e, $e \cdot p \neq 0$. For our purposes, however, we impose an even stronger condition: that p not be equidistant from any two vertices (members of V_{24}). It really does not matter where we select p, and in computer calculations one might just use a random number generator which will select a suitable p with a high probability and then test it to make sure that it does satisfy all required conditions. For this example, however, since we are doing it by hand, we spent some time and selected the point

$$p = (27, 9, 3, 1)$$

in order to make the calculations a little easier. The additional conditions mean that the vertices are totally ordered by their (increasing) distance from p. We call this *Fricke–Klein (FK) order* and observe that it is an extension of stability order. Since

$$
\begin{aligned}
\|v - p\|^2 &= (v - p) \cdot (v - p)\\
&= v \cdot v - 2v \cdot p + p \cdot p\\
&= \|v\|^2 + \|p\|^2 - 2v \cdot p
\end{aligned}
$$

$\forall v, w \in V_{24}$, $v \leq_{FK} w$ iff $p \cdot v \geq p \cdot w$, so it is relatively easy to compute, e.g. $(27, 9, 3, 1) \cdot (1, 1, 0, 0) = 27 + 9 = 36$. This is the largest possible value of $p \cdot v$, so $(1, 1, 0, 0)$ must be the minimum element in the stability order (the unique vertex in the fundamental chamber). V_{24}, sorted into Fricke–Klein order is

$$
\begin{aligned}
&(1,1,0,0)\,,(1,0,1,0)\,,(1,0,0,1)\,,(1,0,0,-1)\,,(1,0,-1,0)\,,\\
&(1,-1,0,0)\,,(0,1,1,0)\,,(0,1,0,1,)\,,(0,1,0,-1,)\,,(0,1,-1,0)\,,\\
&(0,0,1,1)\,,(0,0,1,-1)\,,(0,0,-1,1)\,,(0,0,-1,-1)\,,(0,-1,1,0)\,,\\
&(0,-1,0,1)\,,(0,-1,0,-1)\,,(0,-1,-1,0)\,,(-1,1,0,0)\,,(-1,0,1,0)\,,\\
&(-1,0,0,1)\,,(-1,0,0,-1)\,,(-1,0,-1,0)\,,(-1,-1,0,0)\,.
\end{aligned}
$$

Exercise 5.5 *Is this* $Fricke–Klein$ *order correct?*

Next we must find our Coxeter basis, i.e. the four reflections (out of our list of 24) whose fixed hyperplanes bound the fundamental chamber. Since p is in the fundamental chamber and $p \cdot e$ is the distance from p to the hyperplane orthogonal to e (e is now assumed to be normalized and note that we chose the orientations of the roots so that $p \cdot e > 0$), we sort the normalized roots in increasing order of $p \cdot e$: $(0, 0, 0, 1)$, $\frac{1}{\sqrt{2}}(0, 0, 1, -1)$, $\frac{1}{\sqrt{2}}(0, 0, 1, 1)$, $(0, 0, 1, 0)$, $\frac{1}{\sqrt{2}}(0, 1, -1, 0)$, $\frac{1}{\sqrt{2}}(0, 1, 0, -1)$, $\frac{1}{2}(1, -1, -1, -1)$, $\frac{1}{\sqrt{2}}(0, 1, 0, -1)$, $\frac{1}{2}(1, -1, -1, 1)$, $\frac{1}{\sqrt{2}}(0, 1, 1, 0)$, $(0, 1, 0, 0)$, $\frac{1}{2}(1, -1, 1, -1)$, $\frac{1}{2}(1, -1, 1, 1)$, $\frac{1}{\sqrt{2}}(1, -1, 0, 0)$, $\frac{1}{2}(1, 1, -1, -1)$, $\frac{1}{\sqrt{2}}(1, 0, -1, 0)$, $\frac{1}{2}(1, 1, -1, 1)$, $\frac{1}{2}(1, 1, 1, -1)$, $\frac{1}{\sqrt{2}}(1, 0, 0, -1)$, $\frac{1}{\sqrt{2}}(1, 0, 0, 1)$, $\frac{1}{2}(1, 1, 1, 1)$, $\frac{1}{\sqrt{2}}(1, 0, 1, 0)$, $\frac{1}{\sqrt{2}}(1, 1, 0, 0)$, $(1, 0, 0, 0)$.

Exercise 5.6 *Is this sorting of the roots correct?*

We can assert that $\mathcal{R}_1 = (0, 0, 0, 1)$ is basic because its fixed hyperplane is closest to the Fricke–Klein point, p. In general, having sorted the reflections by the distance of their fixed hyperplanes from p and starting with the closest one, $\mathcal{R}_1$, we can proceed up the list and identify successive basis elements by the fact that if $\mathcal{R}$ is not basic then the perpendicular from p to $H(\mathcal{R})$ will have to pass through a basic hyperplane, $H(\mathcal{R}_i)$, before it gets to $H(\mathcal{R})$. This means that $H(\mathcal{R}_i)$ is closer to p than $H(\mathcal{R})$ and would already have been identified as basic. If $\mathcal{R}_1, \ldots, \mathcal{R}_b$ have already been identified as basic reflections, and $b < d$, then this is equivalent to the equation

$$(p - te) \cdot e_i = 0$$

having a solution, t, $0 < t < p \cdot e$, for some i, $1 \leq i \leq b$.

Theorem 5.9 *$\mathcal{R}$ is basic iff $\forall i$, $1 \leq i \leq b$,*

$$e \cdot e_i \leq \frac{p \cdot e_i}{p \cdot e}.$$

Proof $0 = (p - te) \cdot e_i = p \cdot e_i - te \cdot e_i$. If $e \cdot e_i = 0$ there is no solution. If $e \cdot e_i \neq 0$, the solution is

$$t = \frac{p \cdot e_i}{e \cdot e_i}$$

which is < 0 if $e \cdot e_i < 0$. If $e \cdot e_i > 0$, $\frac{p \cdot e_i}{p \cdot e} = t < p \cdot e$ is equivalent to

$$e \cdot e_i > \frac{p \cdot e_i}{p \cdot e}$$

whose negation is

$$e \cdot e_i \leq \frac{p \cdot e_i}{p \cdot e}.$$

□

The process ends when $b = d$ since $\mathcal{R}_1, ..., \mathcal{R}_d$ is our basis. Applying this to V_{24} we have: $e_1 = (0, 0, 0, 1)$ and $e_2 = \frac{1}{\sqrt{2}}(0, 0, 1, -1)$ since

$$(0, 0, 0, 1) \cdot \frac{1}{\sqrt{2}}(0, 0, 1, -1) = \frac{-1}{\sqrt{2}} < 0.$$

The next candidate, $e = \frac{1}{\sqrt{2}}(0, 0, 1, 1)$ is not basic, however, since $e \cdot e_1 = \frac{1}{\sqrt{2}}$ and

$$\frac{p \cdot e_1}{p \cdot e} = \frac{1}{2\sqrt{2}} < \frac{1}{\sqrt{2}}.$$

Proceeding in this fashion we find that $e_3 = \frac{1}{\sqrt{2}}(0, 1, -1, 0)$, fifth on the list, and $e_4 = \frac{1}{2}(1, -1, -1, -1)$, seventh.

Exercise 5.7 *Verify that $\mathcal{R}_1, \mathcal{R}_2, \mathcal{R}_3, \mathcal{R}_4$ do indeed form a basis since the dot products of their roots, e_1, e_2, e_3, e_4, respectively, satisfy the conditions for the Coxeter graph F_4 (see Table 5.1). Label the vertices of F_4 appropriately.*

Exercise 5.8 *Show that the conditions of Theorem 5.9 will be satisfied, when they are all satisfied, because $e \cdot e_i \leq 0, 1 \leq i \leq b$. This only holds though when we choose the roots so that $p \cdot e > 0$ and then arrange them in Fricke–Klein order.*

5.5.1.1 Generating the stability order

Having a basis and the minimum element, $(1, 1, 0, 0)$, in the stability order of V_{24}, we proceed to generate the rest. Applying $\mathcal{R}_i$, $1 \leq i \leq 4$ to $(1, 1, 0, 0)$ in turn, we find that it is fixed by all except $\mathcal{R}_3$ and

$$\begin{aligned}\mathcal{R}_3\,(1,1,0,0) &= (1,1,0,0) - 2\left[(1,1,0,0)\cdot\frac{1}{\sqrt{2}}\,(0,1,-1,0)\right]\\ &\quad\times\frac{1}{\sqrt{2}}\,(0,1,-1,0)\\ &= (1,1,0,0)-(0,1,-1,0)\\ &= (1,0,1,0)\,.\end{aligned}$$

Repeating the process with $(1, 0, 1, 0)$, ignoring $\mathcal{R}_3$ since that would only take us back to $(1, 1, 0, 0)$, we find that it is fixed by all except $\mathcal{R}_2$, and

$$\mathcal{R}_2\,(1,0,1,0) = (1,0,0,1)\,.$$

$(1, 0, 0, 1)$ is also succeeded by just

$$\mathcal{R}_1\,(1,0,0,1) = (1,0,0,-1)\,.$$

Then it gets a bit more interesting since $(1, 0, 0, -1)$ is succeeded by two vertices in the weak stability order,

$$\mathcal{R}_2\,(1,0,0,-1) = (1,0,-1,0)$$

and

$$\mathcal{R}_4\,(1,0,0,-1) = (0,1,1,0)\,.$$

Proceeding in this manner, we generate the Hasse diagram of the weak stability order, whose edges are solid lines and then the additional edges of the strong stability order (the dashed lines), shown in Fig. 5.8.

5.5.1.2 Generating the derived network

Fig. 5.9 shows the diagram of the derived network of V_{24}. As before, the minimum values of $|\Theta\,(S)|$ for each value of $k = |S|$ are in boldface. The easiest way to compute the weight of each vertex ($|\Theta\,(S)|$) is to calculate the marginal contribution made by one of its maximal elements. For instance, the vertex labled $(0, 1, 0, -1)$ has weight $30 + (8 - 2\cdot 3) = 32$ because its predecessor has weight 30, the degree of each vertex is eight and three of its neighbors, $(0, 1, 1, 0)$, $(1, 0, 0, -1)$, $(1, 0, 0, -1)$, are already in the set (see Lemma 1.1 of Chapter 1).

Exercise 5.9 *Calculate the ratio by which the number of sets to be examined in solving the EIP on V_{24} was reduced by stabilization.*

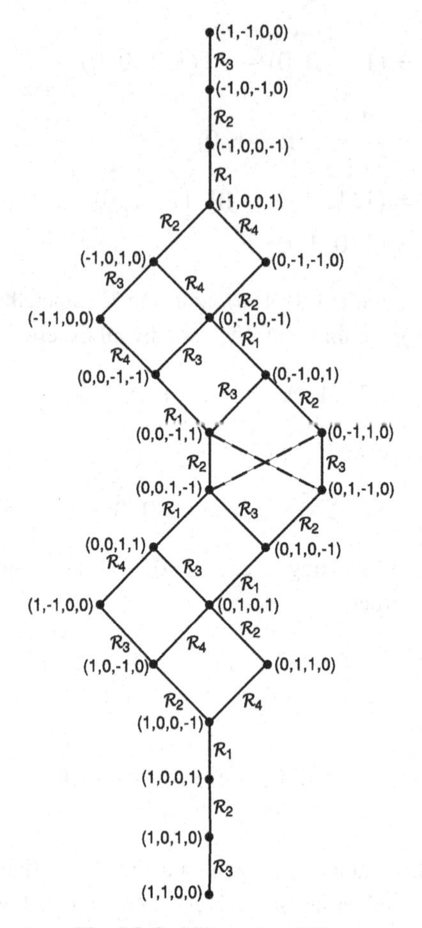

Fig. 5.8 Stability order of V_{24}.

5.5.2 The 120-vertex

The calculation (see [**11**]) that gave the solution of the EIP (and wirelength) for V_{120} was essentially like that given for V_{24} in the previous section, but there were some small differences which we now describe. The coordinates given by Coxeter [**28**] for the vertices of V_{120} are irrational, involving the golden mean, $\tau = \frac{1+\sqrt{5}}{2}$. This complicates their representation in the computer. We could have represented them exactly by using software which will do arithmetic in the field of rationals extended by $\sqrt{5}$, but we chose to use finite decimal approximations of the irrationals involved, do real arithmetic and (since the result of a calculation

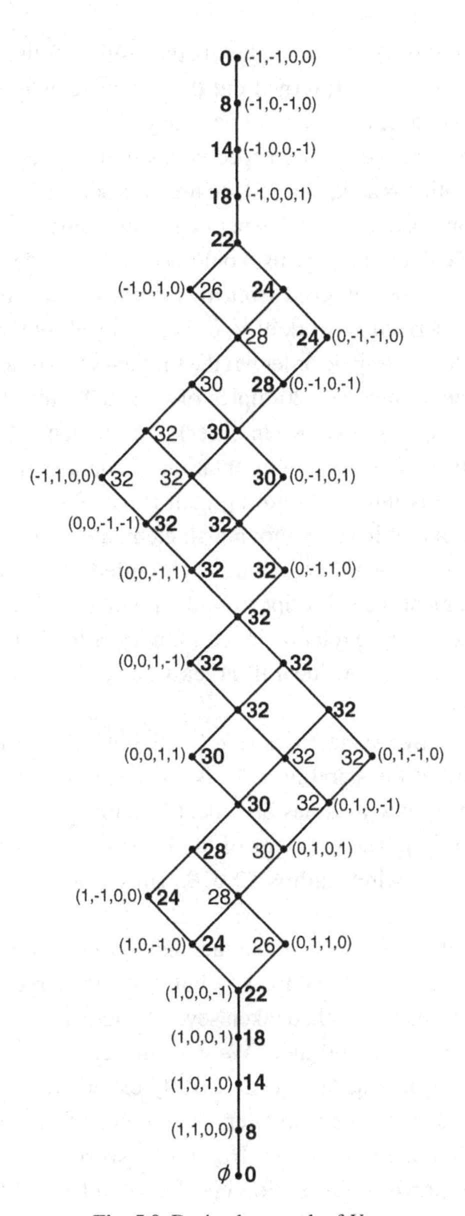

Fig. 5.9 Derived network of V_{24}.

had to be one of a finite set of vectors) identify the result with the candidate which matched it sufficiently well. It turned out that we were able to do everything with the standard six decimal places of accuracy.

Up to this point we have not been specific about the generation of a derived network. The definition is adequate and too much structure can inhibit intuition. Also all the previous examples were generated in an ad hoc fashion with the aim of producing a nice diagram and this would be difficult to describe. However, since computers may be involved from this point on, we must be specific. The vertices of the derived network of V_{120}, i.e. the ideals of the stability order, were generated in lexicographic order wrt the Fricke–Klein order on V_{120}. These "vertices" were represented as 120-tuples of 0s and 1s, the ith entry being 1 iff the ith vertex (wrt to Fricke–Klein order) is in the ideal. Starting with the null set (all 0s), we go from the representation of one ideal to its successor in lexicographic order (assuming it is not V_{120} (all 1s)) by finding the least element not in the ideal, adding it in and removing all elements lower on the list except those below something higher up. All that is required for this operation, if we successively decrement the subscript from the point at which we added in the new element, is the covering relation in the stability order. For each element on the way down, it is in the new ideal iff at least one of its covering elements is already in.

V_{120} has 60 reflective symmetries (see [**28**]), but still just four basic ones. The Hasse diagram of the stability order is too large to represent nicely on a page so we will not show it. It has 883 ideals compared to 33 for the dodecahedron and 41 for V_{24}. The solutions of the EIP for all k, $0 \leq k \leq 120$, sum to 12 616 whereas the wirelength is 12 620. Thus V_{120} does not have nested solutions!

At the time (1979) that we (X. Berenguer and the author) solved the EIP for V_{120}, we attempted to do the same for V_{600} but it was too large. The calculation for V_{120} with its 883 stable sets had taken several hours on the fastest computer we had at UCR, an IBM mainframe. We guestimated that V_{600} had at least a million stable sets which made it seem awfully expensive. In recent years S. Bezrukov suggested that since computers are so much faster now, it might be possible to complete the calculation. He and P. Koch wrote a new program which verified the previous results for V_{120} but it still failed to terminate for V_{600}. Since Berenguer and I had guestimated no more than 32 million stable sets for V_{600} something was wrong. D. Dreier and I redid the calculation and realized what the problem was: the guestimated upper bound had been way too low. Our 500 MHz PC generated a billion stable sets per day for several days with no end in sight. After examining the output, we now guestimate that V_{600} has about 10^{16} stable sets (that is ten million billion!). A big improvement on the $2^{600} \simeq 4.15 \times 10^{180}$ required by brute force, but still beyond our means. We shall return to this problem in Chapter 9.

5.5.3 Cayley graphs of Coxeter groups

We are now in position to give an answer, not a complete answer but a good one, to the question at the end of the introduction to this chapter. What d-dimensional diagram graphs (besides those of the convex regular solids) have vertex-transitive symmetry groups generated by reflections, all of which are stabilizing? Such a graph will have exactly one vertex, v_0, in the fundamental chamber of its Coxeter group, W. Let A be the set of basic reflections which fix v_0. $F(A) = C_0 \cap \bigcap_{\mathcal{R}_i \in A} H(\mathcal{R}_i)$, the face of the fundamental chamber, C_0, which contains v_0, will be a product of faces of the chambers of the irreducible Coxeter subgroups of W. In order for G to be fully d-dimensional, none of those subfaces can be the origin. If that is the case, then the set of all images of v_0 under W will generate a d-dimensional convex polytope. The graph of that convex polytope has W as a group of symmetries and all of its reflections are stabilizing. We believe that the isomorphism type of that convex polytope only depends on $F(A)$, just like the isomorphism type of its stability order (see Theorem 5.6). And just as that stability order was identified with a Bruhat poset, the diagram graph can be identified with the graph whose vertices are the left cosets

$$\{gW_A : g \in W\}$$

and whose edges are the set of pairs

$$\{\{gW_A, g\mathcal{R}_iW_A\} : g \in W \text{ and } 1 \leq i \leq d\}.$$

Exercise 5.10 *Show that this defines a simple (undirected) graph.*

For $A = \emptyset$, $W_A = \mathcal{I}$, the trivial group, our graph is the Cayley graph of W wrt its Coxeter generators. We therefore call this graph the *Coxeter–Cayley graph of W wrt A* and denote it by $CC(W; A)$. If W is reducible, i.e. the Coxeter basis $\Omega = \{\mathcal{R}_1, \mathcal{R}_2, ..., \mathcal{R}_d\}$ can be partitioned into mutually orthogonal subsets, Ω_1 and Ω_2 so $W \simeq W_{\Omega_1} \times W_{\Omega_2}$, then,

$$\begin{aligned} CC(W; A) &= CC\left(W_{\Omega_1} \times W_{\Omega_2}; A \cap \Omega_1 + A \cap \Omega_2\right) \\ &= CC\left(W_{\Omega_1}; A \cap \Omega_1\right) \times CC\left(W_{\Omega_2}; A \cap \Omega_2\right). \end{aligned}$$

Since we have a catalog of all the (finite) irreducible Coxeter groups, we can generate all multisets (subsets with multiplicities) of them whose product will be d-dimensional. Having that product, we can generate all nontrivial subsets, A, of its generators, and compute $CC(W; A)$. Even more generally, each Coxeter generator, $\mathcal{R}_i$, can have a weight, which can be thought of as multiplicity or length, associated with it. If every edge, $e = \{gW_A, g\mathcal{R}_i W_A\}$, is assigned that weight, $w(e) = w(\mathcal{R}_i)$ and the boundary functional is extended to

$$|\Theta(S)| = \sum_{\substack{e \in E \\ \partial(e)=\{u,v\} \\ u \in S, v \notin S}} w(e),$$

and the theory of stabilization still applies. If the weights are nonnegative and sum to 1, they may be thought of as transition probabilities for a random walk on the Cayley graph of W wrt its Coxeter generators, a very interesting mathematical object.

5.6 Extended stabilization

The property which we required for a reflective symmetry of a diagram graph, G, to be stabilizing (Section 3.2.4) was chosen because

(1) It makes $Stab_{\mathcal{R},p}$ a Steiner operation by giving the key inequality

$$\left|\Theta\left(Stab_{\mathcal{R},p}(S)\right)\right| \leq |\Theta(S)|$$

for all $S \subseteq V$.

(2) It is easily verified in a number of interesting cases. In particular we have Theorem 3.1 which shows that all reflective symmetries of convex polytopes are stabilizing.

There remained obvious possibilities for a less restrictive definition which would still give the key inequality but we felt that adding to the complexity of the definition and its verification could only be justified by significant applications. Later, R. J. McEliece came up with just such an application when analyzing the reliability of communications networks. F. Harary had shown [**43**] that the maximum connectivity of any graph with n vertices and $m \leq \binom{n}{2}$ edges is

$$\varkappa(n, m) = \begin{cases} 0 & \text{if } m < n - 1 \\ \lfloor 2m/n \rfloor & \text{if } m \geq n - 1, \end{cases}$$

$2m/n$ being the average degree in such a graph. For $2m/n = \varkappa$, an even integer, Harary showed that the graph $H(n, m)$ whose vertices are $\{0, 1, ..., n - 1\}$ with an edge between i and j if $|i - j \pmod{n}| \leq \varkappa$ has n vertices, m edges and is $\varkappa$-connected.

McEliece proposed to use Harary's graphs as the wiring diagram of a communications network since, given its number of stations (vertices) and communications links (edges), it would require the failure of the maximum number, $\varkappa$, of stations (or links) to disrupt its communications. In analyzing $H(n, m)$ he wished to solve its EIP. Note that the graph of Fig. 5.10 has symmetry group $I_2(8)$ but only half of its eight reflections are stabilizing.

A reflective symmetry, $\mathcal{R}$, of a diagram graph, G, in $\mathbb{R}^d$, is not stabilizing only if there is an edge $e \in E$ with $\partial(e) = \{v, w\}$ with

$$\|v - p\| < \|\mathcal{R}(v) - p\| \text{ and } \|w - p\| > \|\mathcal{R}(w) - p\|.$$

With a dotted line representing (the fixed hyperplane of) $\mathcal{R}$, this gives something like Fig. 5.11 .

By symmetry, G must also contain the edge $\mathcal{R}(e)$ from $v' = \mathcal{R}(v)$ to $w' = \mathcal{R}(w)$ (as in Theorem 3.1) resulting in Fig. 5.12.

As it stands, stabilization wrt $\mathcal{R}$ and p would not be a Steiner operation because $|\Theta(\{v, w\})| = 0$ while $Stab_{\mathcal{R},p}(\{v, w\}) = \{v, w'\}$ and $|\Theta(\{v, w'\})| = 4$. However, if there is an edge between v and w' and its

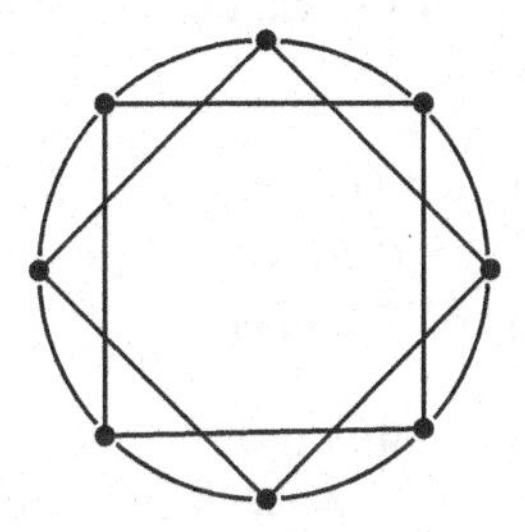

Fig. 5.10 A diagram of Harary's graph $H(8, 16)$.

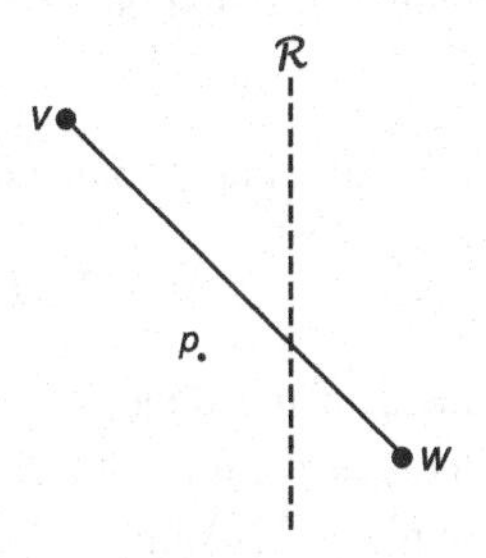

Fig. 5.11

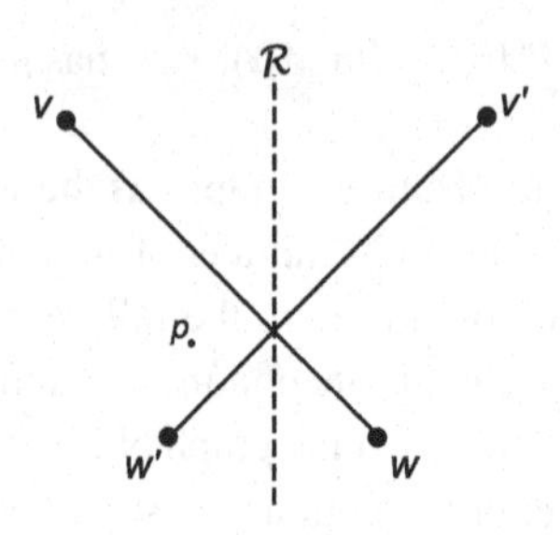

Fig. 5.12

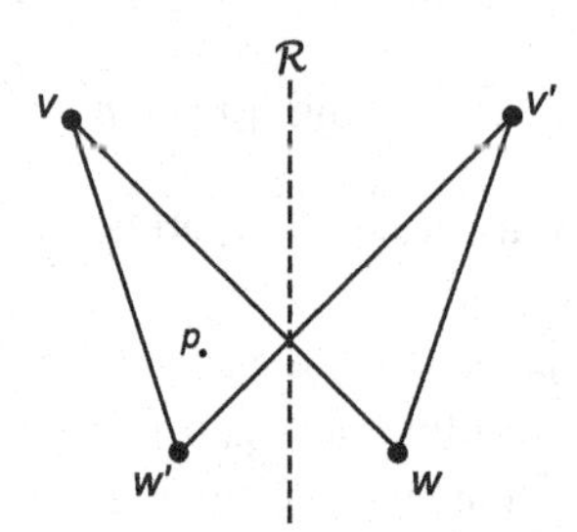

Fig. 5.13 A butterfly.

mirror image, we have Fig. 5.13, which, because of its shape, we call a *butterfly*.

Definition 5.2 Given a diagram G *in* $\mathbb{R}^d$, a reflective symmetry, $\mathcal{R}$, of G, is called *extended stabilizing if* $\forall e \in E$, $\partial(e) = \{v, w\}$ with v and w on opposite sides of the fixed hyperplane of $\mathcal{R}$, either

(1) $\mathcal{R}(e) = e$, i.e. $\mathcal{R}(v) = w$ and $\mathcal{R}(w) = v$, or
(2) $\mathcal{R}(e) = e' \neq e$ so $\mathcal{R}(v) = v' \neq w$, $\mathcal{R}(w) = w' \neq v$ and then $\exists e'' \in E$, $\partial(e'') = \{v, w'\}$, $\partial(\mathcal{R}(e'')) = \{v', w\}$ which give wings to a butterfly.

Theorem 5.10 *If $\mathcal{R}$ is extended stabilizing, then $Stab_{\mathcal{R},p}$ is still a Steiner operation*

Proof Same as for Theorem 3.2 of Chapter 3, except that in showing

$$\left|\Theta\left(Stab_{\mathcal{R},p}(S)\right)\right| \leq |\Theta(S)|$$

we have another case to consider: the edge, e, which is in $\Theta\left(Stab_{\mathcal{R},p}(S)\right)$ but not in $\Theta(S)$ can have $\partial(e) = \{v, w\}$ with $\mathcal{R}(v) \neq w$. We may assume that $|v - p| < |v' - p|$ where $v' = \mathcal{R}(v)$ and $|w - p| > |w' - p|$ where $w' = \mathcal{R}(w)$. Then $v \in Stab_{\mathcal{R},p}(S)$, $w \notin Stab_{\mathcal{R},p}(S)$ and we have the following subcases:

(1) $|S \cap \{v, w, v', w'\}| = 1$ or 3. Then, as in all previous cases, $e' = \mathcal{R}(e) \in \Theta(S)$ but $e' \notin \Theta(Stab_{\mathcal{R},p}(S))$.

(2) $|S \cap \{v, w, v', w'\}| = 2$. Again there are two cases:

   (a) $v, w \in S$. Then $e, e' \in \Theta(Stab_{\mathcal{R},p}(S))$ and $e, e' \notin \Theta(S)$. However, $\mathcal{R}(e), \mathcal{R}(e') \notin \Theta(Stab_{\mathcal{R},p}(S))$ and $\mathcal{R}(e), \mathcal{R}(e') \in \Theta(S)$ which just offset it.

   (b) $v', w' \in S$.

□

Exercise 5.11 *Write out the argument for Case 2(b) of Theorem 5.10.*

5.6.1 The *k*-pather of *G*

One of the nice features of the original definition of stabilization was the ease with which one could show that all the reflective symmetries of many diagram graphs are stabilizing. If the diagram has no edges which intersect then it is stabilizing without the need for butterflies. The extended definition will generally be more difficult to verify but some interesting cases are covered by the following observations.

Given a simple graph, $G = (V, E)$, the *k-pather* of G is the simple graph $G^{(k)} = (V, E^{(k)})$, where

$$E^{(k)} = \{\{v, w\} : \exists \text{ a } v\text{-}w \text{ path of length} \leq k \text{ in } G\}.$$

Note that the adjacency matrix of $G^{(k)}$ is the kth Boolean power of the adjacency matrix of G.

5.6.1.1 Example

Harary's graph $H(n, m)$ is $\mathbb{Z}_n^{(k)}$, the k-pather of the n-cycle.

Theorem 5.11 *If $\mathcal{R}$ is a reflective symmetry of a diagram graph, G, and extended stabilizing, then it is extended stabilizing for $G^{(k)}$, $k = 0, 1, 2, \ldots$*

Proof Proceed by induction. It is trivially true for $k = 0, 1$. Suppose it has been proved for all natural numbers less than $k > 1$ and that we have $\{v, w\} \in E^{(k)}$, $|v - p| < |\mathcal{R}(v) - p|$, $|w - p| > |\mathcal{R}(w) - p|$ and $\mathcal{R}(v) \neq w$. We must show that $\{v, \mathcal{R}(w)\} \in E^{(k)}$. Since $\{v, w\} \in E^{(k)}$, there is a path of length less than or equal to k from v to w in G. Let u be the last vertex before w in that path and consider the following cases: □

(1) If $u = \mathcal{R}(w)$, nothing more need be done, and we have a $(k-1)$-path from v to $\mathcal{R}(w)$.
(2) If $u \neq \mathcal{R}(w)$, then
  (a) If $u = \mathcal{R}(u)$, then add $\mathcal{R}(\{u, w\}) = \{u, \mathcal{R}(w)\}$ to the $(k-1)$-path from v to u.
  (b) If $u \neq \mathcal{R}(u)$, then
    (i) If $\|u - p\| < \|\mathcal{R}(u) - p\|$, then, by the definition of extended stabilizing, $\{u, w\}$ generates a butterfly in G, so $\{u, \mathcal{R}(w)\} \in E$.
    (ii) If $\|u - p\| > \|\mathcal{R}(u) - p\|$, then since $\{v, u\} \in E^{(k-1)}$, by the inductive hypothesis $\{v, \mathcal{R}(u)\} \in E^{(k-1)}$ which means a $(k-1)$-path from v to $\mathcal{R}(u)$. Add the edge $\mathcal{R}(\{u, w\}) = \{\mathcal{R}(u), \mathcal{R}(w)\}$ to it.

In all cases we have constructed a k-path from v to $\mathcal{R}(w)$ in G.

Note that this theorem does not hold for stabilization itself (recall Fig. 5.10 where only half of the eight reflective symmetries of $\mathbb{Z}_8^{(2)}$ are stabilizing).

5.6.2 Applications

5.6.2.1 $\mathbb{Z}_n^{(k)}$, the Harary graph

All of the reflective symmetries of $\mathbb{Z}_n$ are stabilizing and therefore extended stabilizing. By Theorem 5.11 they are extended stabilizing for $\mathbb{Z}_n^{(k)}$ which means that the stability order of $\mathbb{Z}_n^{(k)}$ is at least as large as for $\mathbb{Z}_n$. We observed in Example 3.1 of Section 3.2.7 of Chapter 3 that the stability order of $\mathbb{Z}_n$ is is already total, the stable sets being intervals of adjacent points on $\mathbb{Z}_n$. This solves McEliece's problem.

5.6.2.2 $\mathbb{Q}_d^{(k)}$, the k-pather of the d-cube

Again, the stability order of $\mathbb{Q}_d^{(k)}$ is at least as large as for $\mathbb{Q}_d$. It could be larger: if $k \geq d$, $\mathbb{Q}_d^{(k)}$ is complete so (with a different representation) its stability order could be total. The stability order of $\mathbb{Q}_4$ was diagrammed in Fig. 3.5 of Chapter 3, so it is the same for $\mathbb{Q}_4^{(2)}$ (in this representation) and the graph of its derived network is the same. The values of $|\Theta(S)|$ will differ, however, because $\mathbb{Q}_4^{(2)}$ has more edges than $\mathbb{Q}_4$. The derived network of $\mathbb{Q}_4^{(2)}$ is diagrammed in Fig. 5.14. Note that $\mathbb{Q}_4^{(2)}$ does not have nested solutions.

5.7 Comments

We have solved the EIP on the graphs of all regular solids exept the 600-vertex one in four dimensions. We shall return to that challenge in Chapter 9.

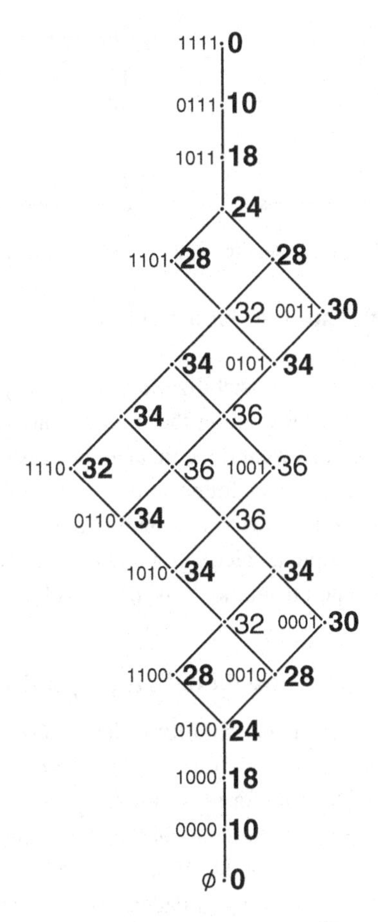

Fig. 5.14 Derived network of $\mathbb{Q}_4^{(2)}$.

5.7.1 Stabilization compared to isomorph rejection

Stabilization, when applicable, is much more powerful than factoring out symmetry. For example, for 2-sets of $\mathbb{Z}_4$, there are two symmetry classes, represented by the sets in Fig. 5.15 but only one stable set. For the 600-vertex there are at least $\frac{2^{600}}{14\,400} \simeq 10^{177}$ symmetry classes of subsets of vertices but only about 10^{16} stable sets. A graph may have a lot of symmetry without any stabilizing symmetry, in which case calculating symmetry classes would appear to be the only choice. But then the problem of finding distinct representatives of the symmetry classes (isomorph rejection) arises. Since this involves the graph

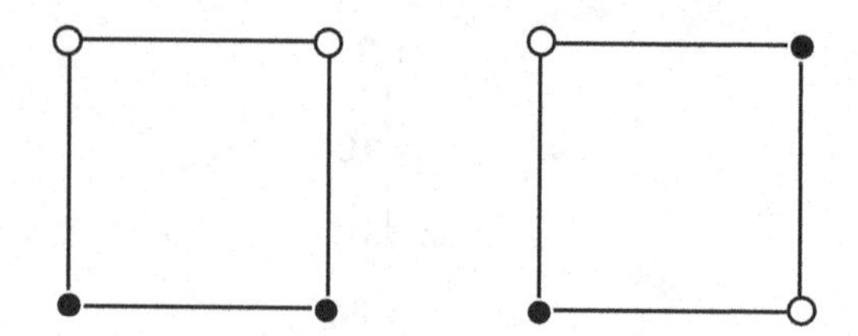

Fig. 5.15 Representitives of symmetry classes in $\mathbb{Z}_4$.

isomorphism problem, known to be difficult, the effort would hardly seem worthwhile.

The computational process of stabilization is also very sturdy. In the 1970s and 1980s, before fully understanding the implications of Coxeter theory for stabilization, the calculations were done in an ad hoc fashion. Years later I was gratified when a student, Joe Vasta, found some missing relations in my stability order for the dodecahedron, showing that he had really understood the theory. It did not mean that the original calculation had been erroneous, just that more vertices could be eliminated from the derived network.

5.7.2 More on deBruijn graphs

Note that the components of the stability order of DB_4 are products of the Bruhat posets of the factors of the Coxeter group, $W = I_2(4) \simeq I_2(2) \times I_2(2)$. We believe this is generally true for reducible Coxeter groups.

DB_4 does not have nested solutions, but the Hamming spheres are solution sets for their cardinalities (0, 1, 5, 11, 15 and 16). Monien and his coworkers at the Center for Parallel Computing in Paderborn have conjectured that this is true for all DB_n and verified it for n up to 8.

6
Higher compression

In this chapter we develop the theory of compression as we did that of stabilization in the previous chapter. We begin with an observation of Bezrukov, that the number of induced edges in products of graphs which have nested solutions for the (induced) EIP is an additive function on ideals in the compressibility order. This observation reduces the EIP on such graphs to a maximum weight ideal problem, a fundamental insight. It facilitates, among other things, the proof of a theorem of Ahlswede and Cai giving a simple sufficient condition for products of graphs to have lexicographic nested solutions for EIP. We present a variety of applications of the Ahlswede–Cai theorem, demonstrating its power and flexibility. Then we present a striking theorem of Bezrukov, Das and Elsässer, that all products of Petersen graphs have nested solutions for EIP but the optimal numberings are not lexicographic.

6.1 Additivity

Definition 6.1 *A set function, $\omega : 2^V \to \mathbb{R}$, is* additive *if $\forall A, B \in 2^V$ such that $A \cap B = \emptyset$, $\omega(A \cup B) = \omega(A) + \omega(B)$.*

Lemma 6.1 *If $\omega : 2^V \to \mathbb{R}$ is additive then*

(1) $\omega(\emptyset) = 0$
(2) $\forall A, B \in 2^V$, $\omega(A \cup B) = \omega(A) + \omega(B) - \omega(A \cap B)$.

Exercise 6.1 *Prove Lemma 6.1.*

Lemma 6.1 is the basis of the Principle of Inclusion–Exclusion, one of the pillars of enumeration and probability theory. It also shows that an additive set function is what lattice-theorists call a *modular function* on the Boolean lattice, $\mathcal{B}_n$ ($= 2^V$, with $n = |V|$). That term has been adopted in much of the literature related to isoperimetric problems (see [**7**] and [**34**]) evidently with the idea that results might be extended to other lattices (as they were for the Principle of Inclusion–Exclusion (see [**85**]). The more promising connection for us, however, is with measure theory, so we prefer to use the terminology (additive function, etc.) of measure theory. If, as is usually the case for our problems, V is finite, then an additive set function is determined by its values on the singleton sets and $\omega(A) = \sum_{a \in A} \omega(\{a\})$.

Lemma 6.2 *On a finite set, an additive set function, $\omega : 2^V \to \mathbb{R}$, is determined by a function $\omega : V \to \mathbb{R}$ and conversely.*

A set function, $\omega : 2^V \to \mathbb{R}$, is *subadditive* (*submodular*) if $\forall A, B \in 2^V$, $\omega(A \cup B) + \omega(A \cap B) \leq \omega(A) + \omega(B)$. It is *superadditive* (*supermodular*) if $\forall A, B \in 2^V$, $\omega(A \cup B) + \omega(A \cap B) \geq \omega(A) + \omega(B)$.

Example 6.1 *$|\Theta(S)|$ is subadditive: For $S, T \subseteq V$, let*

$$[S, T] = \{e \in E : \partial(e) = \{x, y\},\ x \in S \text{ and } y \in T\}$$

and note that $[S, T] = [T, S]$. Then $\Theta(S) = [S, V - S]$ so

$$\begin{aligned}\Theta(S \cup T) &= [S \cup T, V - (S \cup T)] \\ &= [S - T, V - (S \cup T)] + [S \cap T, V - (S \cup T)] \\ &\quad + [T - S, V - (S \cup T)].\end{aligned}$$

and

$$\begin{aligned}\Theta(S \cap T) &= [S \cap T, V - (S \cap T)] \\ &= [S \cap T, S - T] + [S \cap T, T - S] \\ &\quad + [S \cap T, V - (S \cup T)].\end{aligned}$$

Also,

$$\begin{aligned}\Theta(S) &= [S, V - S] \\ &= [S \cap T, V - S] + [S - T, V - S]\end{aligned}$$

and

$$\begin{aligned}\Theta(T) &= [T, V - T] \\ &= [S \cap T, V - T] + [T - S, V - T].\end{aligned}$$

Therefore, the set of edges counted twice in $|\Theta(S \cup T)| + |\Theta(S \cap T)|$ *is*

$$\Theta(S \cup T) \cap \Theta(S \cap T) = [S \cap T, V - (S \cup T)]$$

and the edges counted twice in $|\Theta(S)| + |\Theta(T)|$ *is*

$$\begin{aligned}[S \cap T, V - S] \cap [S \cap T, V - T] &= [S \cap T, (V - S) \cap (V - T)] \\ &= [S \cap T, V - (S \cup T)],\end{aligned}$$

the same set. The set of edges counted just once by $|\Theta(S \cup T)| + |\Theta(S \cap T)|$ *is*

$$\begin{aligned}&[S - T, V - (S \cup T)] + [T - S, V - (S \cup T)] + [S \cap T, S - T] \\ &+ [S \cap T, T - S] = [S - T, V - (S \cup T)] + [T - S, V - (S \cup T)] \\ &+ [S \cap T, (S \cup T) - (S \cap T)],\end{aligned}$$

a subset of the edges,

$$\begin{aligned}&[S - T, V - S] + [T - S, V - T] \\ &+ ([S \cap T, V - S] \cup [S \cap T, V - T] - [S \cap T, V - (S \cup T)]) \\ &= [S - T, V - S] + [T - S, V - T] + [S \cap T, (S \cup T) - (S \cap T)],\end{aligned}$$

counted once by $|\Theta(S)| + |\Theta(T)|$*, so we have our inequality.*

Exercise 6.2 *Prove that* $|\Phi(S)|$ *is subadditive.*

Exercise 6.3 *Prove that* $|E(S)|$ *is superadditive.*

It became apparent over the years that in the course of solution of a number of isoperimetric problems, the objective function, originally sub- or superadditive on 2^V, becomes additive when restricted to the stable or compressed sets, i.e. the ideals of an appropriate poset. For an ideal, S and minimal element $x \in V - S$, this means that

$$\omega(S \cup \{x\}) + \omega(S \cap \{x\}) = \omega(S) + \omega(x).$$

Since

$$\omega(S \cap \{x\}) = \omega(\emptyset) = 0$$

we have

$$\omega(S \cup \{x\}) = \omega(S) + \omega(x).$$

Therefore for all ideals S

$$\omega(S) = \sum_{x \in S} \omega(x).$$

Example 6.2 *VIP on Q_d: For any stable set S (an ideal in the stability order of Q_d), $|\Phi(S)| = \sum_{a \in S} j_0(a)$ where*

$$j_0(a) = \begin{cases} d & \text{if } a = 0^d \\ \min\{j : a_j = 1\} - 2 & \text{if } a \neq 0^d \end{cases}$$

(see Lemma 4.1 of Chapter 4).

Exercise 6.4 *EIP on Q_d: For any stable set S, $|E(S)| = \sum_{a \in S} r(a)$, where $r(a) = \sum_{j=1}^{d} a_j = |\{j : a_j = 1\}|$.*

Finally, Bezrukov [**16**] made the following general observation which began to explain these fortuitous coincidences.

Lemma 6.3 *If graphs G and H have nested solutions for the (induced) EIP, then for all ideals, $S \subseteq V_{G \times H} = V_G \times V_H$ in the compressibility order (i.e. product order, see Example 3.4.2 of Chapter 3),*

$$|E(S)| = \sum_{(i,j) \in S} [\Delta(i) + \Delta(j)]$$

where

$$\Delta(i) = \max_{\substack{S \subseteq V_G \\ |S| = i}} |E(S)| - \max_{\substack{S \subseteq V_G \\ |S| = i-1}} |E(S)|.$$

$$\Delta(j) = \max_{\substack{S \subseteq V_H \\ |S| = j}} |E(S)| - \max_{\substack{S \subseteq V_H \\ |S| = j-1}} |E(S)|.$$

Proof By the definition of a product graph, the only edges incident to (i, j) differ from it in just one component. If, as in the remarks above, S is an ideal in the product order and (i, j) is a minimal element of $V - S$, then the marginal contribution of (i, j) to $|E(S + \{(i, j)\})|$ will be the number of such edges where the vertex at the other end is of the form (g, j) with $g < i$ or (i, h) with $h < j$. These are counted by $\Delta(i)$ and $\Delta(j)$ respectively. □

This result holds for arbitrary products and ideals in any compressibility order (which must be at least as strong as the product order).

6.2 The MWI problem

6.2.1 Definitions

Given a poset $\mathcal{P} = (P, \leq)$ (see Section 3.2.7 of Chapter 3) and $S \subseteq \mathcal{P}$, then let

$$\overleftarrow{S} = \{x \in \mathcal{P} : \exists y \in S \,\&\, x \leq y\}.$$

If $S = \overleftarrow{S}$, then S is an ideal and conversely. Note that $\overleftarrow{\left(\overleftarrow{S}\right)} = \overleftarrow{S}$, so for any $S \subseteq \mathcal{P}$, $\overleftarrow{S}$ is an ideal, *the ideal generated by* S. The set of all ideals of $\mathcal{P}$ (also known as the vertex set of the derived network of $\mathcal{P}$), partially ordered by containment, will be denoted $\mathfrak{I}(\mathcal{P})$. Then the *maximum weight ideal (*MWI*) problem* on $\mathcal{P}$ with weight $\omega : P \to \mathbb{R}$, is to compute

$$\max_{\substack{S \in \mathfrak{I}(\mathcal{P}) \\ |S|=k}} \omega(S),$$

$\forall k \in \mathbb{Z}_+$, where ω has been extended to a set function by additivity.

Dually, let

$$\overrightarrow{S} = \{y \in P : \exists x \in S \,\&\, x \leq y\}.$$

If $S = \overrightarrow{S}$, then S is called a *filter*. Again, $\overrightarrow{\left(\overrightarrow{S}\right)} = \overrightarrow{S}$, so for any $S \subseteq P$, $\overrightarrow{S}$ is a filter, *the filter generated by* S. Also, the complement of a filter is an ideal and vice versa.

Lemma 6.4 *The minimum weight ideal problem is reducible to the maximum weight ideal problem and vice versa so the two are equivalent.*

Proof The complement, $P - S$, of an ideal, $S \in \mathfrak{I}(\mathcal{P})$, is a filter in $\mathcal{P}$ which is an ideal in $\mathcal{P}^*$, the dual of $\mathcal{P}$. The conclusion follows from the fact that $\omega(P - S) = \omega(P) - \omega(S)$. □

Exercise 6.5 *Any MWI problem is equivalent to one with positive weights.*

There is good news and bad news about the MWI problem. The bad news is that there is no polynomial bounded algorithm for it and it is not likely there will ever be one. This follows from (see [**39**] for terminology):

Theorem 6.1 *The MWI problem is NP complete.*

Proof By reduction of max clique: Given a simple graph $G = (V, E, \partial)$ (no loops or multiple edges), we may assume that $E \subseteq \binom{V}{2}$ and ∂ is the identity. Let $\mathcal{P}(G)$ be the set $V \cup E$, partially ordered by $\subseteq$ and weighted by dimension, dim (vertices have weight 0 and edges weight 1). Then G has a k-clique iff

$$\max_{\substack{S\in\mathfrak{I}(\mathcal{P})\\|S|=k+\binom{k}{2}}} \dim(S) = \binom{k}{2}.$$

□

This means that we must be satisfied with a nonpolynomial algorithm such as "brute force" which generates all ideals to find optimal ones. The good news is that there is an efficient way to generate all ideals: given any total extension, $\mathcal{T}$, of the partial order, $\mathcal{P}$, the members of $\mathfrak{I}(\mathcal{P})$ may be recursively generated in lexicographic order (wrt $\mathcal{T}$).

Exercise 6.6 *Compute the Dedekind numbers, $|\mathfrak{I}(\mathcal{B}_n)|$, the number of ideals in the Boolean lattice on n generators, for $n \leq 3$. Hint: generate them lexicographically wrt lex order on $\mathcal{B}_n$.*

If more than one partial order on a set is under consideration, we denote the order relation of $\mathcal{P}$ by $\leq_{\mathcal{P}}$.

Theorem 6.2 *If $G = H_1 \times H_2 \times ... \times H_d$, $d > 1$, is a product of graphs and each of the H_i has nested solutions for the (induced) EIP, then the EIP on G is equivalent to maximizing $\Delta(S)$ over all ideals in the compressibility order with $|S| = k$ where Δ is the additive set function defined by*

$$\Delta(x) = \sum_{i=1}^{d} \Delta_{G_i}(x_i)$$

for $x = (x_1, x_2, ..., x_d)$ and

$$\Delta_G(k) = \max_{\substack{S\subseteq V\\|S|=k}} |E(S)| - \max_{\substack{S\subseteq V\\|S|=k-1}} |E(S)|.$$

Thus the EIP on G is reduced to a maximum weight ideal problem.

S_k, an optimal ideal of cardinality k, is characterized by the fact that for any other k-ideal, S, of $\mathcal{P}$,

(1) $|S_k - S| = |S - S_k|$, and
(2) $\Delta(S_k - S) \geq \Delta(S - S_k)$.
We shall refer to this as *the optimality criterion for MWI*.

Exercise 6.7 *Show that the EIP on any Coxeter–Cayley graph is reducible to an MWI problem.*

Exercise 6.8 *Show that the VIP on a product of graphs having generative nested solutions does not always reduce to an MWI problem.*

Exercise 6.9 *Show that the VIP on any Coxeter–Cayley graph is reducible to an MWI problem.*

6.3 The Ahlswede–Cai theorem

The full power of compression only emerges when every product of graphs taken from some basic set, $\{H_1, H_2, \ldots\}$ with arbitrary mulitiplicities, has nested solutions. We may then prove that $G = H_{i_1} \times H_{i_2} \times \ldots \times H_{i_d}$ has nested solutions by induction on d. For large d there are many factorizations of G, each giving us the opportunity to apply the inductive hypothesis and strengthen the compressibility order. Such arguments fall naturally into three cases:

- $d = 1$: Compression does not apply.
- $d = 2$: Compression applies, giving the product order on $H_{i_1} \times H_{i_2}$ as compressibility order.
- $d \geq 3$: Compression applies and the compressibility order is generally much stronger than product order on $H_{i_1} \times H_{i_2} \times \ldots \times H_{i_d}$.

Paradoxically, the cases $d = 1, 2$ are often the most difficult. The next theorem, by Ahlswede and Cai [**3**], gives some indication of why this is.

Recall (Chapter 1, Section 1.2.3) that if $\{T_i\}_{i=1}^{n}$ is a sequence of totally ordered sets, then *lexicographic order* on their product, $T_1 \times T_2 \times \ldots \times T_n$, is the total order defined by $x < y$ if $\exists m$ such that $x_1 = y_1, x_2 = y_2, \ldots, x_{m-1} = y_{m-1}$ and $x_m < y_m$ (see also Chapter 3, Section 3.5).

Theorem 6.3 *[**3**] Let Ω be a totally ordered set and $\{H_\alpha : \alpha \in \Omega\}$ be a set of graphs indexed by Ω. If*

(1) $\forall \alpha \in \Omega$, H_α has nested solutions for the (induced) EIP, *and*
(2) $\forall \alpha \leq \beta$, $H_\alpha \times H_\beta$ has lexicographic nested solutions (wrt the total orders for H_α and H_β),

then $\forall \alpha_1 \leq \alpha_2 \leq \ldots \leq \alpha_d$, $H_{\alpha_1} \times H_{\alpha_2} \times \ldots \times H_{\alpha_d}$ has lexicographic nested solutions.

Proof It is assumed true for $d = 1, 2$. Applying the optimality criterion above when $d = 2$ we see that if S_k is the initial segment of lexicographic order of size k, and S is any other ideal of $H_\alpha \times H_\beta$ such that $\exists l, \emptyset \neq S - S_k \subsetneq \{l+1\} \times H_\beta$

and $S_k - S \subset \{l\} \times H_\beta$, then

$$\Delta(S_k - S) \geq \Delta(S - S_k).$$

Now note that the compressibility order, $\mathcal{C}(H_{\alpha_1} \times H_{\alpha_2} \times ... \times H_{\alpha_d})$, contains the chains $\{m\} \times H_{\alpha_2} \times ... \times H_{\alpha_d}$, $1 \leq m \leq |H_{\alpha_1}|$ and that $\forall m$, the top $|H_{\alpha_d}|$ elements of the chain, $(m, |H_{\alpha_2}|, ..., |H_{\alpha_{d-1}}|, l)$, $1 \leq l \leq |H_{\alpha_d}|$, are each covered by $(m+1, 1, ..., 1, l)$, one of the bottom $|H_{\alpha_1}|$ elements of the next chain up. So if S is an ideal in $\mathcal{C}(H_{\alpha_1} \times H_{\alpha_2} \times ... \times H_{\alpha_d})$ of size k which is not S_k, the initial lexicographic segment in $H_{\alpha_1} \times H_{\alpha_2} \times ... \times H_{\alpha_d}$ of the same size, there must exist m, $1 \leq m \leq |H_{\alpha_1}|$, such that $l_0 = \max\{l : (m+1, 1, ..., 1, l) \in S\}$, $1 < l_0 < |H_{\alpha_d}|$ and $k_0 = \max\{(m, |H_{\alpha_2}|, ..., |H_{\alpha_{d-1}}|, l) \in S_k\}$, $l_0 < k_0 < |H_{\alpha_d}|$. If $k_0 + l_0 \leq |H_{\alpha_d}|$ then

$$S - S_k = \{(m+1, 1, ..., 1, l) : 1 \leq l \leq l_0\}$$

and

$$S_k - S = \{(m, |H_{\alpha_2}|, ..., |H_{\alpha_{d-1}}|, l) : k_0 < l \leq l_0 + k_0\}.$$

On the other hand, if $k_0 + l_0 > |H_{\alpha_d}|$ then

$$S - S_k = \{(m+1, 1, ..., 1, l) : k_0 + l_0 - |H_{\alpha_d}| < l \leq l_0\}$$

and

$$S_k - S = \{(m, |H_{\alpha_2}|, ..., |H_{\alpha_{d-1}}|, l) : k_0 < l \leq |H_{\alpha_d}|\}.$$

Let $(S_k - S)'$ be the projection of $S_k - S$ into $H_{\alpha_1} \times H_{\alpha_d}$ and similarly for $(S - S_k)'$. Then

$$\begin{aligned}
\Delta(S_k - S) &\geq \Delta((S_k - S)'), \text{ since } \Delta(|H_{\alpha_s}|) \geq 0, \\
&\geq \Delta((S - S_k)'), \text{ by the optimality criterion,} \\
&= \Delta(S - S_k), \text{ since } \Delta(1) = 0.
\end{aligned}$$

Therefore, by the optimality criterion again, S_k is optimal. □

6.3.1 Applications

In order to apply the Ahlswede–Cai Theorem to the products of a family of basic graphs, we must first show that the basic graphs have nested solutions. Then we must show that the pairwise products of these basic graphs have nested solutions given by lexicographic order wrt the total orders which give solutions on the components. Solving the EIP on an irreducible graph is, in general, a difficult problem and one which we must take on a case-by-case basis. Pairwise products also present a challenge, but since they have been reduced to MWI problems on the product of two total orders, we have considerably more structure with which to work.

Before presenting our concrete applications, we develop some general tools for the two-dimensional cases. We call them *relative compression, relative stabilization* and *elevation*. The following definitions and arguments were constructed with the aid of diagrams and it is suggested that readers construct their own to aid in following them.

(1) *Relative compression* is essentially induction on the length of the component total orders. The idea is quite simple and but let us take a moment to analyze and explore its possibilities. We are trying to solve the MWI problem on an $n_1 \times n_2$ product of total orders and considering some ideal, S. If we have $(i_0, j_0) < (i_1, j_1)$ such that the subrectangle $\{(i_0, j_0) \leq (i, j) \leq (i_1, j_1)\}$ has weights for which the solution of the MWI problem is known to be lexicographic, then we can transform that part of S intersecting the subrectangle to lex order if
   (a) $(i_1 + 1, j_0) \notin S$, and
   (b) either $i_0 = 1$ or $(i_0 - 1, j_1) \in S$.

   Calling the transformed set $RelComp\,(S)$ we may verify that it is an ideal of the same cardinality as S and that $\omega\,(RelComp\,(S)) \geq \omega\,(S)$. It is the extremality of lex order, i.e. the minimality of the first coordinates and the maximality of the second, which makes these conditions sufficient as well as necessary for defining a Steiner operation, independent of the number of members of S in the subrectangle.
(2) *Relative stabilization*: Given an ideal S and (i_0, j_0), if $\forall (i, j) \in S$ such that $(i, j) < (i_0, j_0)$, $i - j > k_0 = i_0 - j_0$ and $j \geq j_0$, $\omega\,(i, j) \leq \omega\,(j + k, i - k)$, and if
   (a) either $n_1 - k_0 \leq n_2$ or $(n_2 + k_0 + 1, j_0) \notin S$, and
   (b) either $i_0 = 1$ or $(i_0 - 1, n_2,) \in S$.

   then any $(i, j) \in S$ such that $i - j > k_0$, $j \geq j_0$, $(i, i - k_0) \notin S$ and $(j + k_0, i - k_0) \notin S$ can be removed and replaced by $(j + k_0, i - k_0)$. Call the resulting set $RelStab\,(S)$. From this definition it follows that $RelStab\,(S)$ is an ideal of the same cardinality as S and that $\omega\,(RelStab\,(S)) \geq \omega\,(S)$.
(3) *Elevation*: If $S = RelStab\,(S)$ above then we can define another Steiner operation called *elevation*. For j $\geq j_0$, let $I\,(j) = \max\,\{i : (i, j) \in S\}$. If $I\,(j) > \overline{j} + k_0$, where $\overline{j} = \max\left\{j' : I\left(j'\right) = I\,(j)\right\}$, then let $J\,(j) =$

$\max\{j' : (j + k_0, j') \in S\}$. If $J(j) < n_2$ then remove $(I(j), j)$ from S and replace it by $(j + k, J(j) + 1)$. Call the set $Elev(S)$. Again we observe that $Elev(S)$ is an ideal of the same cardinality as S and $\omega(Elev(S)) \geq \omega(S)$. Elevation may be repeated several times, each time peeling off the top element, $(J(j), j)$, of the jth row, $j_0 \leq j < j_1$ if $I(j) \geq i_1$.

6.3.1.1 Products of complete graphs

Theorem 6.4 *(Lindsey [73]) The EIP on $K_{n_1} \times K_{n_2} \times ... \times K_{n_d}$, a product of complete graphs with $n_1 \leq n_2 \leq ... \leq n_d$, has lexicographic nested solutions.*

Proof By the Ahlswede–Cai Theorem, we need only show it for $d = 1, 2$. $d = 1$ is trivial (see Example 1.2.1 of Chapter 1); $d = 2$, however, is not. If we identify the compressibility order on $K_{n_1} \times K_{n_2}$ with

$$\{(i, j) : 1 \leq i \leq n_1 \ \& \ 1 \leq j \leq n_2\}$$

ordered coordinatewise, then its marginal weight is

$$\Delta_{n_1,n_2}(i, j) = (i - 1) + (j - 1),$$

which is the rank of (i, j). We proceed by induction on n_1. For $n_1 = 1$ the theorem is equivalent to the $m = 1$ case already proved. Assume it true for $n_1 - 1 \geq 1$. Suppose the theorem is false and that S is an optimal ideal in $K_{n_1} \times K_{n_2}$, minimal wrt lex order. Let a be the maximum member of S (wrt lex order) and let b be the minimum member of $V - S$. Then, since S is not an initial segment of lex order, $a >_{lex} b$ and we have the following cases:

(1) If $a_1 < n_1$, then relative compression wrt $(1, 1)$ and $(n_1 - 1, n_2)$ gives a contradiction since S is contained in the interval between those two elements which is isomorphic to $K_{n_1-1} \times K_{n_2}$.
(2) If $b_1 > 1$, a similar contradiction is given by relative compression wrt $(1, 2)$ and (n_1, n_2). The interval is isomorphic to $K_{n_1-1} \times K_{n_2}$ with the weights increased by 1.
(3) If $a_1 = n_1$ and $b_1 = 1$, then relative stabilization and elevation with respect to $(1, \max\{a_2 - n_1 + 2, 1\})$ contradicts the minimality of S wrt lex order.

□

Corollary 6.1 *Lexicographic order on $K_{n_1} \times K_{n_2} \times ... \times K_{n_m}$ with $n_1 \geq n_2 \geq ... \geq n_m$ solves its wirelength problem.*

Exercise 6.10 *Show that the product $K_n \times G$, G any graph with nested solutions for the EIP and $|V_G| \leq n$, has lex nested solutions.*

6.3.1.2 Products of complete bipartite graphs

Ahlswede and Cai's original application of their theorem was the following:

Theorem 6.5 *[3] The EIP on $K_{n_1,n_1} \times K_{n_2,n_2} \times ... \times K_{n_m,n_m}$, a product of complete bipartite graphs with $n_1 \leq n_2 \leq ... \leq n_m$, has lexicographic nested solutions.*

Proof By the Ahlswede–Cai theorem, we need only show it for $m = 1, 2$. The case $m = 1$ is easily solved since for any $S \subseteq V_{K_{n,n}} = A + B$

$$|E(S)| = |S \cap A|\,|S \cap B|,$$

and

$$|S| = |S \cap A| + |S \cap B|.$$

Therefore S should be split as evenly as possible between A and B to maximize $|E(S)|$. Any numbering of $V_{K_{n,n}}$ which assigns odd numbers to A and even numbers to B will have initial segments which are solutions. From this we may calculate the marginal weight for $K_{n,n}$:

$$\Delta_n(i) = \left\lfloor \frac{i}{2} \right\rfloor$$

for $i = 1, 2, ..., 2n$. Then the marginal weight for the compressibility order of $K_{n_1,n_1} \times K_{n_2,n_2}$ is

$$\Delta_{n_1,n_2}(i, j) = \Delta_{n_1}(i) + \Delta_{n_2}(j).$$

As in the preceding proof, we proceed by induction on n_1. For $n_1 = 1$ it follows from the monotonicity of $\Delta_{n_1}(i)$ and the fact that

$$\begin{aligned}\Delta_{2,n_2}(2, 2n_2 - 1) &= 1 + (n_2 - 1) = n_2 \\ &= \Delta_{2,n_2}(1, 2n_2)\end{aligned}$$

and $\forall j < 2n_2 - 1$,

$$\begin{aligned}\Delta_{n_1,n_2}(2, j) + \Delta_{n_1,n_2}(2, j+1) &= \left(1 + \left\lfloor \frac{j}{2} \right\rfloor\right) + \left(1 + \left\lfloor \frac{j+1}{2} \right\rfloor\right) \\ &= \left\lfloor \frac{j+2}{2} \right\rfloor + \left\lfloor \frac{j+3}{2} \right\rfloor \\ &= \Delta_{n_1,n_2}(1, j+2) + \Delta_{n_1,n_2}(1, j+3).\end{aligned}$$

So assume it true for $n_1 - 1 \geq 1$ and suppose that S is a lexicographically minimal optimal ideal in $K_{n_1,n_1} \times K_{n_2,n_2}$ but not an initial segment of lex order.

Let a be the maximum member of S (wrt lex order) and let b be the minimum member of $V - S$. Then $a >_{lex} b$ and we have the following cases:

(1) If $a_1 \leq 2(n_1 - 1)$, we obtain a contradiction by relative compression wrt $(1, 1)$ and $(2(n_1 - 1), n_2)$ since S is contained in the interval between those two elements which is isomorphic to $K_{n_1-1,n_1-1} \times K_{n_2,n_2}$.
(2) If $b_1 > 2$, we again obtain a contradiction by relative compression wrt $(3, 1)$ and $(2n_1, 2n_2)$. The interval is isomorphic to $K_{n_1-1,n_1-1} \times K_{n_2,n_2}$ with the weights increased by 1.
(3) If $a_1 \geq 2n_1 - 1$ and $b_1 \leq 2$, then S must be fixed by relative stabilization wrt $(1, 1)$, $(1, 3)$ $(1, 5)$, ... The only way that this can happen is if $b_1 = 2$, so $(1, 2n_2) \in S$, $a_1 = 2n_1$ or $2n_1 - 1$ so $(2n_1 - 1, 1) \in S$ and $(3, 2) \notin S$. But then S is not fixed by relative stabilization and elevation wrt $(2, 1)$, a contradiction.

□

6.3.1.3 Products of crosspolytopes

Theorem 6.6 *The EIP on* $\maltese_{n_1} \times \maltese_{n_2} \times \ldots \times \maltese_{n_m}$*, a product of crosspolytopes with* $n_1 \leq n_2 \leq \ldots \leq n_m$*, has lexicographic nested solutions.*

Proof By the Ahlswede–Cai theorem, we need only show it for $m = 1, 2$. The case $m = 1$ has already been proved in Example 2 of Section 3.2.7 of Chapter 3. From this we may calculate the marginal weight, Δ_n, for $\maltese_n$:

$$\Delta_n(i) = \begin{cases} i - 1 & \text{if } 1 \leq i \leq n, \\ i - 2 & \text{if } n + 1 \leq i \leq 2n. \end{cases}$$

If we identify the compressibility order on $\maltese_{n_1} \times \maltese_{n_2}$ with

$$\{(i, j) : 1 \leq i \leq 2n_1 \ \& \ 1 \leq j \leq 2n_2\}$$

ordered coordinatewise, then its marginal weight is

$$\Delta_{n_1,n_2}(i, j) = \Delta_{n_1}(i) + \Delta_{n_2}(j).$$

As in the preceding proofs, we proceed by induction on n_1. For $n_1 = 1$, $\Delta_{n_1}(i) = 0$ and the theorem follows from the monotonicity of $\Delta_{n_2}(j)$. Assume it true for $n_1 - 1 \geq 1$. If S is an optimal ideal in $\maltese_{n_1} \times \maltese_{n_2}$ which is minimal wrt lex order but not an initial segment, let a be the maximum member of S (wrt lex order) and let b be the minimum member of $V - S$. Then $a >_{lex} b$ and we have the following cases:

(1) If $a_1 < 2n_1$ and $b_1 > 1$, then the result follows by relative compression wrt $(2, 1)$ and $(2n_1 - 1, 2n_2)$ since the interval between those two elements is isomorphic to $\maltese_{n_1-1} \times \maltese_{n_2}$, and its weight just differs by 1.

(2) If $b_1 = 1$, S must be fixed by relative stabilization and elevation with respect to $(1, n_2 - n_1 + 1)$ up to $(1, 2n_2 - n_1 + 1)$. This can only be true if

$$\begin{aligned}&\{(i, j) \in S : j \geq n_2 - n_1 + 1\} \\ &\quad = \{(1, n_2 - n_1 + 1), (1, n_2 - n_1 + 2), ..., (1, b_2 - 1)\}.\end{aligned}$$

S must then also be fixed by relative stabilization and elevation wrt $(1, 1)$ up to $(1, n_2 - n_1)$, which means that

$$S = \{(1, 1), (1, 2), ..., (1, b_2 - 1)\},$$

an initial segment of lex order.

(3) If $a_1 = 2n_1$, the result follows from the previous case by duality (complementation). □

Exercise 6.11 *Show that all products of small cycles ($\mathbb{Z}_n$, $n \leq 4$) have nested solutions.*

6.4 The Bezrukov–Das–Elsässer theorem

6.4.1 The Petersen graph and its products

Denote by P the *Petersen graph (diagrammed in Fig. 6.1)*. The Petersen graph is a favorite with combinatorialists because of its unusual and extremal properties. It has the reputation of being a universal counterexample, the rock upon which many a pretty conjecture has foundered. It is the logo of the *Journal of Graph Theory* and there is even a book devoted to its lore [**55**].

Let $P^d = P \times P \times ... \times P$, the d-fold product of Petersen graphs. Bezrukov, Elsässer & Das[**18**], motivated by computer scientists interested in using P^d as the connection graph for multiprocessing computers, solved the EIP for P^d. Their solution and proof, which we now turn to, is a beautiful illustration of the power of compression and may well point the way for future developments.

6.4.2 The solution for $d = 1$

Lemma 6.5 *The Petersen graph has nested solutions, maximizing $|E(S)|$, the number of induced edges, over all $S \subseteq V$ with $|S| = k$. The initial segments of the numbering, $\mathcal{P}_1$, shown in Fig. 6.1 are such optimal sets.*

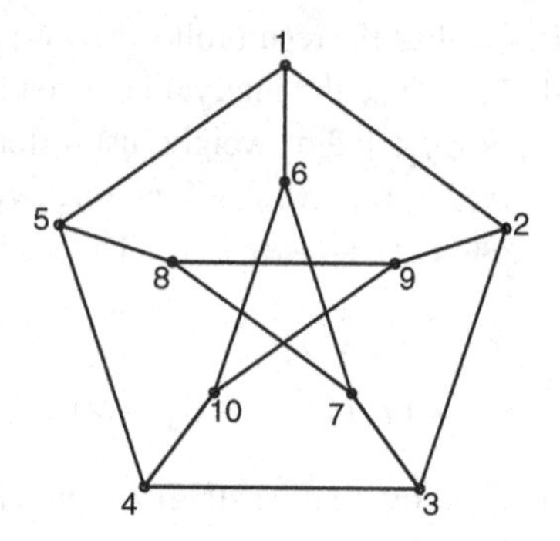

Fig. 6.1 The Petersen graph.

Proof P has girth (the minimum length of a circuit) of 5. Thus for $0 \le k < 5$, solution sets induce trees, and for $k = 5$, a circuit (see Section 1.2.2 of Chapter 1). For $5 < k \le 10$, the result follows from that for $10 - k$ by complementation since (see Section 1.2 of Chapter 1)

$$\begin{aligned}|E(V-S)| &= \frac{1}{2}\left[\delta k - |\Theta(V-S)|\right], \text{ where } \delta = \delta(P) = 3,\\ &= \frac{1}{2}\left[\delta k - |\Theta(S)|\right], \text{ since } \Theta(V-S) = \Theta(S),\\ &= \frac{1}{2}\left[\delta k - \left[\delta(10-k) - 2|E(S)|\right]\right]\\ &= \delta(k-5) + |E(S)|.\end{aligned}$$

□

6.4.3 The solution for $d = 2$

Since our problem has nested solutions for $d = 1$, the theory of compression applies and we need only consider ideals in the compressibility order $\mathcal{C} = \mathcal{P}_1 \times \mathcal{P}_1$, on P^2.

Corollary 6.2 *For a compressed set S of P^d (i.e. an ideal in $\mathcal{C}$ and therefore $\mathcal{P}_1^d$),*

$$|E_{P^d}(S)| = \sum_{a \in S} \sum_{j=1}^{d} \Delta_{\mathcal{P}_1}(a_j),$$

where the vertices of P have been identified with the integers 1, 2, ..., 9, 10 *by* $\mathcal{P}_1$ *(See Fig. 6.1).*

The values of $\Delta_{\mathcal{P}_1}$ are given in the following table:

i	0	1	2	3	4	5	6	7	8	9	10
$\min_{\lvert S\rvert=i} \lvert E_P(S)\rvert$	0	0	1	2	3	5	6	8	10	12	15
$\Delta_{\mathcal{P}_1}(i)$	0	0	1	1	1	2	1	2	2	2	3

and the resulting weighting of $\mathcal{P}_1 \times \mathcal{P}_1$ is:

a_2										
10	3	4	4	4	5	4	5	5	5	6
9	2	3	3	3	4	3	4	4	4	5
8	2	3	3	3	4	3	4	4	4	5
7	2	3	3	3	4	3	4	4	4	5
6	1	2	2	2	3	2	3	3	3	4
5	2	3	3	3	4	3	4	4	4	5
4	1	2	2	2	3	2	3	3	3	4
3	1	2	2	2	3	2	3	3	3	4
2	1	2	2	2	3	2	3	3	3	4
1	0	1	1	1	2	1	2	2	2	3
Δ	1	2	3	4	5	6	7	8	9	10
	a_1									

From the previous table, assuming that P^2 has nested solutions, we can find a numbering which would give such solutions by maximizing the weight of each successive entry:

a_2										
10	19	20	30	49	50	69	70	80	99	100
9	17	18	29	47	48	67	68	79	97	98
8	15	16	28	45	46	65	66	78	95	96
7	13	14	27	43	44	63	64	77	93	94
6	11	12	26	41	42	61	62	76	91	92
5	9	10	25	39	40	59	60	75	89	90
4	7	8	24	37	38	57	58	74	87	88
3	5	6	23	35	36	55	56	73	85	86
2	3	4	22	33	34	53	54	72	83	84
1	1	2	21	31	32	51	52	71	81	82
$\mathcal{P}_2$	1	2	3	4	5	6	7	8	9	10
	a_1									

This total ordering of P^2, which we call $\mathcal{P}_2$, has been proven to have initial segments which maximize $\lvert E(S)\rvert$ for all ideals of the same cardinality. The original proof [18] was by computer, calculating $|E(S)|$ for all $\binom{20}{10} = 184\,756$ ideals of the compressibility order, $\mathcal{P}_1^2$. In Chapter 9 however, we shall present a proof which can be checked by hand.

6.4.4 Solution for all $d > 2$

From evidence such as this, one may follow Bezrukov, Das and Elsässer in guessing that the following recursively defined total order, $\mathcal{P}_d$, on V_{P^d}, will give solutions:

$$(a_1, ..., a_{d-1}, a_d) >_{\mathcal{P}_d} (b_1, ..., b_{d-1}, b_d) \text{ iff}$$

(1) $a_d - 1 > b_d$, or
(2) $a_d - 1 = b_d$ and $b_d \in \{2, 3, 5, 7, 8\}$, or
(3) $a_d - 1 = b_d$ and $b_d \in \{1, 4, 6, 9\}$ and $(a_1, ..., a_{d-1}) \geq_{\mathcal{P}_{d-1}} (b_1, ..., b_{d-1})$, or
(4) $a_d = b_d$ and $(a_1, ..., a_{d-1}) >_{\mathcal{P}_{d-1}} (b_1, ..., b_{d-1})$, or
(5) $a_d + 1 = b_d$ and $b_d \in \{2, 5, 7, 10\}$ and $(a_1, ..., a_{d-1}) >_{\mathcal{P}_{d-1}} (b_1, ..., b_{d-1})$.

Lemma 6.6 *$\mathcal{P}_d$ is a total order on V_{P^d}.*

Proof Note that parts (3) and (5) of the definition say that if $|a_d - b_d| = 1$ and the smaller of a_d, b_d is in the set $\{1, 4, 6, 9\}$ then the order of a and b is determined by their remaining entries. Thus for every pair $a \neq b$, either $a >_{\mathcal{P}_d} b$ or $b >_{\mathcal{P}_d} a$. It remains to show that $\mathcal{P}_d$ is transitive and this we prove by induction on d. We have already proven it for $d = 1, 2$, so let $d > 2$. Assume that $a >_{\mathcal{P}_d} b$ by Case i above, that $b >_{\mathcal{P}_d} c$ by Case j and consider the $5 \times 5 = 25$ combinations, (i, j), by which these two inequalities could happen:

$(1, 1), (1, 2), (1, 3), (1, 4)$: $a_d - 1 > b_d \geq c_d$ so $a >_{\mathcal{P}_d} c$ by part (1).
$(1, 5)$: Same as $(1, 1)$, etc. *unless* $a_d = b_d + 2 = (c_d - 1) + 2 = c_d + 1$, but then $c_d \in \{2, 5, 7, 10\}$ and $c_d \neq 10$ implies $c_d \in \{2, 3, 5, 7, 8\}$ so $a >_{\mathcal{P}_d} c$ by part (2).
$(2, 1), (2, 2), (2, 3)$: Same as $(1, 1)$, etc.
$(2, 4)$: $a >_{\mathcal{P}_d} c$ by part (2).
$(2, 5)$: $a_d = b_d + 1 = (c_d - 1) + 1 = c_d$, but $a_d \in \{3, 4, 6, 8, 9\}$ & $c_d \in \{2, 5, 7, 10\}$ which is impossible.
$(3, 1), (3, 2), (3, 3)$: Same as $(1, 1)$, etc.

(3, 4): $a_d - 1 = b_d = c_d$ and $(a_1, ..., a_{d-1}) \geq_{\mathcal{P}_{d-1}} (b_1, ..., b_{d-1}) >_{\mathcal{P}_{d-1}} (c_1, ..., c_{d-1})$. By the inductive hypothesis $(a_1, ..., a_{d-1}) >_{\mathcal{P}_{d-1}} (c_1, ..., c_{d-1})$ so $a >_{\mathcal{P}_d} c$ by part (3).

(3, 5): Like (2, 5), $a_d = c_d \in \{2, 5, 7, 10\}$ but

$$(a_1, ..., a_{d-1}) \geq {\mathcal{P}_{d-1}} (b_1, ..., b_{d-1}) \\ > {\mathcal{P}_{d-1}} (c_1, ..., c_{d-1})$$

and by the inductive hypothesis, $(a_1, ..., a_{d-1}) >_{\mathcal{P}_{d-1}} (c_1, ..., c_{d-1})$ so $a >_{\mathcal{P}_d} c$ by part (4).

(4, 1): Same as (1, 1), etc.

(4, 2): $a_d - 1 = b_d = c_d$ and $c_d = b_d \in \{2, 3, 5, 7, 8\}$ so by part (2) $a >_{\mathcal{P}_d} c$.

(4, 3): Same as (3, 4).

(4, 4): $a_d = b_d = c_d$ and $(a_1, ..., a_{d-1}) >_{\mathcal{P}_{d-1}} (b_1, ..., b_{d-1}) >_{\mathcal{P}_{d-1}} (c_1, ..., c_{d-1})$ so by the inductive hypothesis $(a_1, ..., a_{d-1}) >_{\mathcal{P}_{d-1}} (c_1, ..., c_{d-1})$ and $a >_{\mathcal{P}_d} c$ by part (4).

(4, 5): $a_d + 1 = b_d + 1 = c_d$ and

$$(a_1, ..., a_{d-1}) > {\mathcal{P}_{d-1}} (b_1, ..., b_{d-1}) \\ > {\mathcal{P}_{d-1}} (c_1, ..., c_{d-1})$$

so by the inductive hypothesis $(a_1, ..., a_{d-1}) >_{\mathcal{P}_{d-1}} (c_1, ..., c_{d-1})$ and $a >_{\mathcal{P}_d} c$ by part (5).

(5, 1): Same as (1, 1), etc. unless $b_d = c_d + 2$ so $a_d = b_d - 1 = (c_d + 2) - 1 = c_d + 1$, but then $b_d \in \{2, 5, 7, 10\}$, so $c_d \in \{3, 5, 8\} \subset \{2, 3, 5, 7, 8\}$ and again $a >_{\mathcal{P}_d} c$ by part (2).

(5, 2): $a_d = b_d - 1 = (c_d + 1) - 1 = c_d$ but $a_d \in \{1, 4, 6, 9\}$ and $c_d \in \{2, 3, 5, 7, 8\}$ which is impossible.

(5, 3): $a_d = c_d$ as in (5, 2) but

$$(a_1, ..., a_{d-1}) \geq {\mathcal{P}_{d-1}} (b_1, ..., b_{d-1}) \\ > {\mathcal{P}_{d-1}} (c_1, ..., c_{d-1})$$

so by the inductive hypothesis $(a_1, ..., a_{d-1}) >_{\mathcal{P}_{d-1}} (c_1, ..., c_{d-1})$ and $a >_{\mathcal{P}_d} c$ by part (4).

(5, 4): $a_d + 1 = b_d = c_d$ and $c_d = b_d \in \{2, 5, 7, 10\}$ and

$$(a_1, ..., a_{d-1}) \geq {\mathcal{P}_{d-1}} (b_1, ..., b_{d-1}) \\ > {\mathcal{P}_{d-1}} (c_1, ..., c_{d-1})$$

so by the inductive hypothesis $(a_1, ..., a_{d-1}) >_{\mathcal{P}_{d-1}} (c_1, ..., c_{d-1})$. Thus $a >_{\mathcal{P}_d} c$ by part (5).

(5, 5): $b_d \in \{2, 5, 7, 10\}$ and $c_d \in \{2, 5, 7, 10\}$ but $b_d = c_d + 1$ which is impossible.

That concludes our proof. □

Since $\mathcal{P}_d$ is a total order on P^d, let $s(x)$ be the successor of x in that total order.

Lemma 6.7 *If* $(x_1, ..., x_{d-1}, x_d) \neq (10, ..., 10, 10)$ *then*

$$s(x_1, ..., x_{d-1}, x_d)$$

$$= \begin{cases} (s(x_1, ..., x_{d-1}), x_d) & \text{if } x_d \in \{3, 8\} \text{ and } (x_1, ..., x_{d-1}) \neq (10, ..., 10) \\ (s(x_1, ..., x_{d-1}), x_d - 1) & \text{if } x_d \in \{2, 5, 7, 10\} \text{ and } (x_1, ..., x_{d-1}) \neq (10, ..., 10) \\ (1, ..., 1, x_d + 1) & \text{if } x_d \in \{2, 3, 5, 7, 8\} \text{ and } (x_1, ..., x_{d-1}) \neq (10, ..., 10) \\ (x_1, ..., x_{d-1}, x_d + 1) & \text{if } x_d \in \{1, 4, 6, 9\}. \end{cases}$$

Proof Call the right-hand side of the equation in our lemma $s'(x)$, so we are trying to prove $\forall x \neq (10)^d$, $s(x) = s'(x)$. From the definition of the total order, $\mathcal{P}_d$, we see that $\forall x \neq (10)^d$, $x <_{\mathcal{P}_d} s'(x)$, so we need only show that $s'(x)$ is the least of those elements which are greater than x, i.e. $x <_{\mathcal{P}_d} y \Rightarrow s'(x) \leq_{\mathcal{P}_d} y$. Again we get down to cases, examining the cases by which $s'(x)$ and $x <_{\mathcal{P}_d} y$ are defined:

Case A. $x_d \in \{3, 8\}$ and $(x_1, ..., x_{d-1}) \neq (10)^{d-1}$ so $s'(x) = (s(x_1, ..., x_{d-1}), x_d)$:

1. $y_d > x_d + 1 = x'_d + 1 \Rightarrow s'(x) <_{\mathcal{P}_d} y$ by (1).
2. $y_d = x_d + 1 = x'_d + 1 \Rightarrow s'(x) \leq_{\mathcal{P}_d} y$ by (2) or (3).
3. $y_d = x_d + 1$ and $x_d \in \{1, 4, 6, 9\}$ does not occur.
4. $y_d = x_d = x'_d$ and $(x_1, ..., x_{d-1}) <_{\mathcal{P}_{d-1}} (y_1, ..., y_{d-1})$
   $\Rightarrow s(x_1, ..., x_{d-1}) \leq_{\mathcal{P}_{d-1}} (y_1, ..., y_{d-1})$
   $\Rightarrow s'(x) \leq_{\mathcal{P}_d} y$ by (4).
5. $y_d + 1 = x_d$ and $x_d \in \{2, 5, 7, 10\}$ does not occur.

Case B. $x_d \in \{2, 5, 7, 10\}$ and $(x_1, ..., x_{d-1}) \neq (10)^{d-1}$ so $s'(x) = (s(x_1, ..., x_{d-1}), x_d - 1)$:

1. $y_d > x_d + 1 = x'_d + 2 \Rightarrow s'(x) <_{\mathcal{P}_d} y$ by 1.
2. Same as B.1.
3. Same as B.1.
4. $y_d = x_d$ and $(x_1, ..., x_{d-1}) <_{\mathcal{P}_{d-1}} (y_1, ..., y_{d-1})$
   $\Rightarrow s(x_1, ..., x_{d-1}) \leq_{\mathcal{P}_{d-1}} (y_1, ..., y_{d-1})$
   $\Rightarrow s'(x) \leq_{\mathcal{P}_d} y$ by (3) since $x'_d = x_d - 1 \in \{1, 4, 6, 9\}$.
5. $y_d = x_d - 1 = x'_d$ and $(x_1, ..., x_{d-1}) <_{\mathcal{P}_{d-1}} (y_1, ..., y_{d-1})$
   $\Rightarrow s(x_1, ..., x_{d-1}) \leq_{\mathcal{P}_{d-1}} (y_1, ..., y_{d-1})$
   $\Rightarrow s'(x) \leq_{\mathcal{P}_d} y$.

Case C. $x_d \in \{2, 3, 5, 7, 8\}$ and $(x_{1,\dots,} x_{d-1}) = (10)^{d-1}$ so $s'(x) = (1, \dots, 1, x_d + 1)$:

1. $y_d > x_d + 1 = x'_d \Rightarrow s'(x) <_{\mathcal{P}_d} y$ by 1, 2 or 3 since $(1, \dots, 1) \leq_{\mathcal{P}_{d-1}} (y_1, \dots, y_{d-1})$.
2. $y_d = x_d + 1 = x'_d$ $\Rightarrow s'(x) \leq_{\mathcal{P}_d} y$ by (4) since $(1, \dots, 1) \leq_{\mathcal{P}_{d-1}} (y_1, \dots, y_{d-1})$.
3. Same as C.2.
4. $y_d = x_d$ and $(x_1, \dots, x_{d-1}) <_{\mathcal{P}_{d-1}} (y_1, \dots, y_{d-1})$ but this is impossible since $(x_{1,\dots,} x_{d-1}) = (10, \dots, 10)$.
5. Same as C.4.

Case D. $x_d \in \{1, 4, 6, 9\}$ so $s'(x) = (x_{1,\dots,} x_{d-1}, x_d + 1)$

1. $y_d > x_d + 1 = x'_d \in \{2, 5, 7,$ (but not 10)$\}$ $\Rightarrow s'(x) \leq_{\mathcal{P}_d} y$ by (1) or (2).
2. $x_d \in \{2, 3, 5, 7, 8\}$ does not occur.
3. $y_d = x_d + 1 = x'_d$ and $(x'_{1,\dots,} x'_{d-1}) = (x_{1,\dots,} x_{d-1})$ $\leq_{\mathcal{P}_{d-1}} (y_1, \dots, y_{d-1}) \Rightarrow s'(x) \leq_{\mathcal{P}_d} y$ by (4).
4. $y_d = x_d = x'_d - 1$ and $(x'_{1,\dots,} x'_{d-1}) = (x_{1,\dots,} x_{d-1})$ $\leq_{\mathcal{P}_{d-1}} (y_1, \dots, y_{d-1}) \Rightarrow x'_d \in \{2, 5, 7, 10\}$ $\Rightarrow s'(x) \leq_{\mathcal{P}_d} y$ by (5).
5. $x_d \in \{2, 5, 7, 10\}$ does not occur.

□

Theorem 6.7 *$\forall d \geq 1$ and $\forall k, 0 \leq k \leq 10^d$, S_k, the initial k-segment of $\mathcal{P}_d$, maximizes $|E(S)|$ over all $S \subseteq V_{\mathcal{P}_d}$ with $|S| = k$.*

Proof We have already proven it for $d = 1, 2$, so we need only consider $d > 2$. Given a compressed set $S \neq S_k$, let a be the maximal member of S and b the minimal member of $V - S$. The $a >_{\mathcal{P}_d} b$. We once more consider the ways that this can happen and show that if it does we can alter S to S', a compressed set closer to S_k ($S' = S - a + b$ in most cases) with $|S'| = |S|$ and $|E(S')| \geq |E(S)|$.

(1) $a_d - 1 > b_d$:

(a) $a_d - b_d \geq 4$: Then $b_d \leq 6$ so
$a = (a_1, \dots, a_{d-1}, a_d) \geq_{\mathcal{C}} (a_1, \dots, a_{d-1}, b_d + 4)$, since
$\mathcal{P}_1^d \subseteq \mathcal{C}, >_{\mathcal{C}} (b_1, \dots b_{d-2}, a_{d-1}, b_d + 2)$, by the definitions of
$\mathcal{P}_d$ & $\mathcal{C}, >_{\mathcal{C}} (b_1, \dots b_{d-2}, b_{d-1}, b_d) = b$.
Therefore, $a >_{\mathcal{C}} b$, a contradiction.

(b) $a_i > 2$ for some i, $1 < i < d$: Then

$(a_1, ..., a_{i-1}, a_i) >_{\mathcal{P}i} (b_1, ...b_{i-1}, 2)$ by (1) or (2). Thus

$$\begin{aligned} a &= (a_1, ..., a_{d-1}, a_d) \\ &\geq_{\mathcal{C}} (b_1, ..., b_{i-1}, 2, a_{i+1}, ..., a_d), \text{ by (4)}, \\ &>_{\mathcal{C}} (b_1, ..., b_{d-1}, b_d) = b, \text{ by (1) and the definition of } \mathcal{C}. \end{aligned}$$

So, we have our contradiction $a >_{\mathcal{C}} b$ again.

(c) $b_i < 9$ for some i, $1 < i < d$: Then $(a_i, a_{i+1}, ..., a_d) >_{\mathcal{P}d-i+1} (9, b_{i+1}, ..., b_d)$, by (1). Thus

$$\begin{aligned} a &= (a_1, ..., a_{d-1}, a_d) \\ &\geq_{\mathcal{C}} (a_1, ..., a_{i-1}, 9, b_{i+1}, ..., b_d), \text{ by definition of } \mathcal{C}, \\ &>_{\mathcal{C}} (b_1, ..., b_{i-1}, b_i, ..., b_d) = b, \text{ by (4), (1) and the definition of } \mathcal{C}. \end{aligned}$$

Again we have our contradiction.

(d) $a_i \leq 2$ and $b_i \geq 9$ for all i, $1 < i < d$ and $a_1 \geq b_1$: Then

$$\begin{aligned} a &= (a_1, ..., a_{d-1}, a_d) \\ &>_{\mathcal{C}} (b_1, ..., a_{d-1}, a_d) \\ &>_{\mathcal{C}} (b_1, ..., b_{d-1}, b_d) = b, \text{ by (1), contradiction.} \end{aligned}$$

(e) $a_i \leq 2$ and $b_i \geq 9$ for all i, $1 < i < d$ and $a_1 < b_1$: Then

$$\begin{aligned} \Delta(b) - \Delta(a) &= \Delta_{\mathcal{P}_1}(b_1) - \Delta_{\mathcal{P}_1}(a_1) \\ &\quad + \textstyle\sum_{1<i<d} \Delta_{\mathcal{P}_1}(b_i) - \Delta_{\mathcal{P}_1}(a_i) \\ &\quad + \left(\Delta_{\mathcal{P}_1}(b_d) - \Delta_{\mathcal{P}_1}(a_d)\right) \\ &\geq (-1) + (d-2) + (-1) \\ &= d - 4. \end{aligned}$$

Thus, our strategy works for all $d > 3$.

(f) $d = 3$: $\Delta(b) - \Delta(a) \geq 0$ above if $a_2 = 1$, $b_2 = 10$, $b_3 \neq 6$ or $a_3 \neq 5$. So we need only consider $a = (a_1, 2, 5)$, $b = (b_1, 9, 6)$ with $a_1 - b_1 \in \{2, 3\}$. Then since $(a_1, 1, 5) <_{\mathcal{C}} (a_1, 2, 5) \in S$ and $(b_1, 10, 6) >_{\mathcal{C}} (b_1, 9, 6) \notin S$ let $S' = S - \{(a_1, 1, 5), (a_1, 2, 5)\} + \{(b_1, 9, 6), (b_1, 10, 6)\}$. $|S'| = |S|$ and $|E(S')| - |E| \geq 2(-1) + (2+3) \geq 0$. S' may not be compressed but we may compress it to obtain a compressed set of the same cardinality and at least as many induced edges. Compression will not displace b or any of its predecessors in S so we are finished with Case 1.

(2) $a_d - 1 = b_d$ and $b_d \in \{2, 3, 5, 7, 8\}$: The analysis of this case is essentially the same as Cases 1 (b)–(e). The only difference is that we now have $\Delta_{\mathcal{P}_1}(b_d) \geq \Delta_{\mathcal{P}_1}(a_d)$ so in subcase (e), $\Delta(b) - \Delta(a) \geq d - 3 \geq 0$ for all $d \geq 3$.

(3) $a_d - 1 = b_d$ and $b_d \in \{1, 4, 6, 9\}$ and $(a_1, ..., a_{d-1}) \geq_{\mathcal{P}_{d-1}} (b_1, ..., b_{d-1})$: But then

$$\begin{aligned} a = (a_1, ..., a_{d-1}, a_d) &\geq_{\mathcal{C}} (b_1, ..., b_{d-1}, a_d) \\ &>_{\mathcal{C}} (b_1, ..., b_{d-1}, b_d) = b, \end{aligned}$$

and we have our contradiction again.

(4) $a_d = b_d$ and $(a_1, ..., a_{d-1}) >_{\mathcal{P}_{d-1}} (b_1, ..., b_{d-1})$: Even more directly, $a >_{\mathcal{C}} b$.

(5) $a_d + 1 = b_d$ and $b_d \in \{2, 5, 7, 10\}$ and $(a_1, ..., a_{d-1}) >_{\mathcal{P}_{d-1}} (b_1, ..., b_{d-1})$:

(a) $b_{d-1} \in \{1, 4, 6, 9\}$: Then $s(b_1, ..., b_{d-1}) = (b_1, ..., b_{d-1} + 1)$ by Lemma 6.7. Therefore

$$\begin{aligned} a &= (a_1, ..., a_{d-1}, a_d) \\ &\geq_{\mathcal{C}} (s(b_1, ..., b_{d-1}), a_d), \text{ since } (a_1, ..., a_{d-1}) \geq_{\mathcal{P}_{d-1}} \\ &\quad (b_1, ..., b_{d-1}), = (b_1, ..., b_{d-1} + 1, a_d) \\ &\quad >_{\mathcal{C}} (b_1, ..., b_{d-1}, a_d + 1) = b, \end{aligned}$$

since $(b_{d-1} + 1, a_d) >_{\mathcal{P}_2} (b_{d-1}, a_d + 1)$.

(b) $b_{d-1} \in \{3, 8\}$:

(i) $(b_1, ..., b_{d-2}) \neq (10, ..., 10)$: Then

$s(b_1, ..., b_{d-1}) = (s(b_1, ..., b_{d-2}), b_{d-1})$ by Lemma 6.7 so

$$\begin{aligned} a = (a_1, ..., a_{d-1}, a_d) &\geq_{\mathcal{C}} (s(b_1, ..., b_{d-2}), b_{d-1}, a_d), \\ &\text{since } (a_1, ..., a_{d-1}) \geq_{\mathcal{P}_{d-1}} s(b_1, ..., b_{d-1}), \\ &>_{\mathcal{C}} (b_1, ..., b_{d-1}, a_d + 1) = b, \end{aligned}$$

since $(s(b_1, ..., b_{d-2}), a_d) >_{\mathcal{P}_{d-1}} (b_1, ..., b_{d-2}, a_d + 1)$, by (5).

(ii) $(b_1, ..., b_{d-2}) = (10, ..., 10)$: Then

$s(b_1, ..., b_{d-1}) = (1, ..., 1, b_{d-1} + 1)$ and

$s(s(b_1, ..., b_{d-1})) = (1, ..., 1, b_{d-1} + 2)$, by Lemma 6.7.

(α) $(a_1, ..., a_{d-1}) >_{\mathcal{P}_{d-1}} s(b_1, ..., b_{d-1})$: Then

$$\begin{aligned} a &= (a_1, ..., a_{d-1}, a_d) \\ &\geq_{\mathcal{C}} (s(s(b_1, ..., b_{d-1})), a_d), \text{ by (3)}, \\ &= (1, ..., 1, b_{d-1} + 2, a_d) >_{\mathcal{C}} (1, ..., 1, b_{d-1} + 1, a_d + 1) \\ &\quad >_{\mathcal{C}} (10, ..., 10, b_{d-1}, a_d + 1) = b. \end{aligned}$$

(β) $(a_1, ..., a_{d-1}) = s(b_1, ..., b_{d-1}) = (1, ..., 1, b_{d-1} + 1)$: Then

$$\begin{aligned} b = (10, ..., 10, b_{d-1}, b_d), \; a &= (1, ..., 1, b_{d-1} + 1, b_d - 1) \\ \text{and } \Delta(b) - \Delta(a) &= \textstyle\sum_{i=1}^{d-2} \Delta_{\mathcal{P}_1}(b_i) - \Delta_{\mathcal{P}_1}(a_i) \\ &\quad + \left(\Delta_{\mathcal{P}_1}(b_{d-1}) - \Delta_{\mathcal{P}_1}(a_{d-1})\right) \\ &\quad + \left(\Delta_{\mathcal{P}_1}(b_d) - \Delta_{\mathcal{P}_1}(a_d)\right) \\ &\geq 3(d-2) + 0 + 1 \\ &= 3(d-2) + 1 \geq 1. \end{aligned}$$

(c) $b_{d-1} \in \{2, 5, 7, 10\}$:

(i) $(b_1, ..., b_{d-2}) = (10, ..., 10)$: Then $b_2 \neq 10$ since $(a_1, ..., a_{d-1}) >_{\mathcal{P}_{d-1}} (b_1, ..., b_{d-1})$. Thus $s(b_1, ..., b_{d-2}, b_{d-1}) = (1, ..., 1, b_{d-1} + 1)$.

(α) $b_2 = 5$: So $b_2 + 1 = 6$ and

$$s\,(s\,(b_1,\ldots,b_{d-2},b_{d-1})) = s\,((1,\ldots,1,b_{d-1}+1))$$
$$= (1,\ldots,1,b_{d-1}+2),$$ by Lemma 6.7, and we can finish the argument as in 5.b.ii.β.

(β) $b_2 \in \{2,7\}$: Then

$s\,((1,\ldots,1,b_{d-1}+1)) = (2,\ldots,1,b_{d-1}+1)$, by Lemma 6.7, and again there are two possibilities. However, having run out of alphabets to index, we just list them.

If $(a_1,\ldots,a_{d-1}) >_{\mathcal{P}_{d-1}} s\,(b_1,\ldots,b_{d-1})$ then

$$\begin{aligned} a &= (a_1,\ldots,a_{d-1},a_d) \\ &\geq_{\mathcal{C}} (s\,(s\,(b_1,\ldots,b_{d-1}))\,,a_d), \text{ by (3)}, \\ &= (2,\ldots,1,b_{d-1}+1,a_d) \\ &>_{\mathcal{C}} (1,\ldots,1,b_{d-1}+1,a_d+1) \\ &>_{\mathcal{C}} (10,\ldots,10,b_{d-1},a_d+1) = b. \end{aligned}$$

Or if $(a_1,\ldots,a_{d-2},a_{d-1}) = s\,(b_1,\ldots,b_{d-2},b_{d-1})$ $= (1,\ldots,1,b_{d-1}+1)$ then

$b = (10,\ldots,10,b_{d-1},b_d)$,

$a = (1,\ldots,1,b_{d-1}+1,b_d-1)$ and,

as in 5.b.ii.β, $\Delta\,(b) - \Delta\,(a) > 0$.

(ii) $(b_1,\ldots,b_{d-2}) \neq (10,\ldots,10)$: Then by Lemma 6.7

$$s\,(b_1,\ldots,b_{d-2},b_{d-1}) = \left(s\left(b_1,\ldots,b_{d-2}\right),b_{d-1}-1\right)$$

and we have our usual two possibilities.

(α) $(a_1,\ldots,a_{d-1}) >_{\mathcal{P}_{d-1}} s\,(b_1,\ldots,b_{d-1})$: Then

$$\begin{aligned} a &= (a_1,\ldots,a_{d-1},a_d) \\ &\geq_{\mathcal{C}} (s\,(s\,(b_1,\ldots,b_{d-1}))\,,a_d), \text{ by (3)}, \\ &= \left(s\left(s\left(b_1,\ldots,b_{d-2}\right),b_{d-1}-1\right),a_d\right) \\ &>_{\mathcal{C}} \left(s\left(b_1,\ldots,b_{d-2}\right),b_{d-1}-1,a_d+1\right), \text{ by (5)}, \\ &>_{\mathcal{C}} \left(b_1,\ldots,b_{d-2},b_{d-1},a_d+1\right) = b, \text{ by (5) again}. \end{aligned}$$

(β) $(a_1,\ldots,a_{d-1}) = s\,(b_1,\ldots,b_{d-1})$
$= \left(s\left(b_1,\ldots,b_{d-2}\right),b_{d-1}-1\right)$:

Then we reapply the analysis of Case 5 to $(a_1,\ldots,a_{d-1})\,,(b_1,\ldots,b_{d-1})$. Either we will replace a with b without decreasing $|E\,(S)|$ or one possibility of dimension $d-2$ will remains. Ultimately we are left with $b = (b_1,b_2,\ldots,b_{d-1},b_d)$, $b_i \in \{2,5,7,10\}\,,\forall i$, $1 \leq i < d$. Also $b_d \neq 10$, and $a = s\,(b) =$ $(b_1+1,b_2-1,\ldots,b_{d-1}-1,b_d-1)$. But then

$$\begin{aligned} \Delta\,(b) - \Delta\,(a) &= \Delta_{\mathcal{P}_1}(b_1) - \Delta_{\mathcal{P}_1}(a_1) \\ &\quad + \textstyle\sum_{i=2}^{d} \Delta_{\mathcal{P}_1}(b_i) - \Delta_{\mathcal{P}_1}(a_i) \\ &\geq -1 + (d-1) = d-2 \geq 0, \end{aligned}$$

and we are done. □

6.5 Comments

Bezrukov and Elsässer [20] showed that powers of the graphs $K_{2n} - (I_1 + I_2 + ... + I_m)$, where $\{I_j\}$ is a parallel family of disjoint complete matchings in K_{2n} and $m \leq n/2$, have lex nested solutions for EIP. The graph of $\maltese_d$ is isomorphic to $K_{2d} - I_1$, so this result is related to our Theorem 6.6. Their proof is another application of the Ahlswede–Cai theorem (Theorem 6.3), but the result has broader significance because it gives an isoperimetric inequality on all regular graphs with the same number of vertices and edges. They state similar results for powers of bipartite graphs $K_{n,n} - (I_1 + I_2 + ... + I_m)$, $m \leq n/2$.

T. Carlson [25] used the logic of Bezrukov, Elsässer and Das to show that powers of $\mathbb{Z}_5$ have nested solutions. The optimal numbering is again not lexicographic. Since $\mathbb{Z}_5$ is a face of the dodecahedron (V_{20}) and the Petersen graph is the quotient of V_{20} by its antipodal symmetry, the obvious next question to consider in this direction is whether the powers of V_{20} have nested solutions for the EIP? A recent computer calculation by Bezrukov has shown that it does not.

Exercise 6.12 *Does the pairwise product of icosahedra, $V_{12} \times V_{12}$, have nested solutions for the EIP?*

$\mathbb{Z}_m \times \mathbb{Z}_n$, $m, n \geq 6$, does not have nested solutions, nor does a product of paths $P_m \times P_n$, $m, n \geq 6$, However, Bollobas and Leader [22] showed that the induced EIP on the product of paths does have nested solutions (the product of paths is not regular so the two variants of EIP are not equivalent).

Can the Ahlswede–Cai theorem be extended to prove the above theorems of Bezrukov–Elsässer–Das and Bollobas–Leader? Is it even true that nested solutions for EIP on all pairwise products from some basic set of graphs implies nested solutions for all products?

The arguments of this chapter, particularly that of the Bezrukov–Das–Elsässer Theorem are "combinatorial" in the old, pejorative sense of that word, i.e. they involve delicate inductions with cases, subcases and subsubcases. It could be very helpful to have an analog of Coxeter theory (in its support role for stabilization) for compression.

Is there an analog of the Ahlswede–Cai theorem for VIP? So far, the results in this direction seem negative. There is no analog of Lemma 6.3, which shows that the EIP on a product of graphs having nested solutions can be reduced to an MWI problem. That may seem puzzling since we observed that there is such a weighting on the stability order of Q_d (Example 6.1.3) but later we shall show that the result may be generalized in a different direction. There is a theorem of Chvátalová [31] and Moghadom [79] which plays much the same role for the VIP that Lindsey's theorem plays for the EIP. They showed that the product of paths $P_{n_1} \times P_{n_2} \times ... \times P_{n_m}$ have generative nested solutions, the optimal numbering being a natural extension of Hales numbering of Q_d. The eight-year gap between Chvátalová's result for $d = 2$ and Moghadam's for $d > 2$ is an indication of the qualitative difference in compression arguments between those two cases. Essentially the same result carries over to products of cycles, $\mathbb{Z}_{n_1} \times \mathbb{Z}_{n_2} \times ... \times \mathbb{Z}_{n_m}$. The products of complete graphs, $K_{n_1} \times K_{n_2} \times ... \times K_{n_m}$, also known as Hamming graphs, do not generally have nested solutions for the VIP. Thus the two families of products of regular graphs, K_n and $\mathbb{Z}_n$, switch roles between the EIP and VIP. We shall return to these unsolved problems in Chapter 10. There are also intriguing analogies between the optimal numberings for the EIP on $K_{n_1} \times K_{n_2} \times ... \times K_{n_m}$ and the VIP on $\mathbb{Z}_{n_1} \times \mathbb{Z}_{n_2} \times ... \times \mathbb{Z}_{n_m}$. The former interpolate balls in the L_∞-norm and the latter interpolate balls in the L_1-norm, suggesting a kind of duality, but nothing has come of it so far.

The problem of finding an optimal uniform 2-partition for the vertices of a graph was mentioned in the Comments to Chapter 1, as graph bisection. Since it is NP-complete ([**39**], p. 210), graph bisection is in some sense equivalent to the full EIP. Also, none of the methods which have been applied to graph bisection depend on the fact that $k = |S| = \left\lfloor \frac{|V|}{2} \right\rfloor$ and extend immediately to EIP. However, the converse of that statement is not true: compression, arguably the most effective tool for obtaining solutions of EIP, depends explicitly on having a nested family of solutions to the EIP, one for each value of k, $0 \leq k \leq |V|$.

E. C. Posner conjectured Corollary 6.1 around 1964. The author considered taking up the challenge but, in light of his experience with Theorem 1.1 of Chapter 1, was daunted by its apparent complexity. J. E. Lindsey, however, succeeded in writing down a proof of Posner's conjecture [**73**], and was awarded the annual prize for outstanding research by an undergraduate at Cal Tech.

A recent paper by Azizoğlu and Eğecioğlu [**8**] contains an interesting variant of Lindsey's theorem (Theorem 6.4). They weight the edges of K_n by

$$\gamma(n) = \begin{cases} \frac{1}{n^2} & \text{if } n \text{ is even,} \\ \frac{1}{n^2-1} & \text{if } n \text{ is odd.} \end{cases}$$

and consider the EIP wrt those weights. The marginal weight on $K_{n_1} \times K_{n_2} \times ... \times K_{n_m}$ is then

$$\Delta(i_1, i_2, ..., i_m) = \sum_{j=1}^{m} \gamma(n_j)(i_j - 1).$$

Theorem 6.8 *[8] The EIP on $K_{n_1} \times K_{n_2} \times \ldots \times K_{n_m}$, a product of complete graphs (weighted as above) with $n_1 \leq n_2 \leq \ldots \leq n_m$, has reverse lex nested solutions.*

The only difference between this and Theorem 6.4 is that the lexicographic order is taken wrt the reverse order of components. The Ahlswede–Cai theorem still applies and the proof of the two-dimensional case is mainly by relative compression. Azizoğlu and Eğecioğlu apply their theorem to compute the isoperimetric number (of a graph, $G = (V, E, \partial)$),

$$i(G) = \min_{\substack{S \subset V \\ |S| \leq |V|/2}} \frac{|\Theta(S)|}{|S|}$$

for a product of paths, $P_{n_1} \times P_{n_2} \times \ldots \times P_{n_m}$. Their result is that

$$i\left(P_{n_1} \times P_{n_2} \times \ldots \times P_{n_m}\right) = \frac{1}{\lfloor n_m/2 \rfloor}$$

where $n_m = \max_{1 \leq j \leq m} n_j$.

7

Isoperimetric problems on infinite graphs

Why infinite graphs? The EIP, or any of its variants, would not seem suited to infinite graphs. On finite graphs we can always find a solution by brute force, evaluating $|\Theta(S)|$ for all $2^{|V|}$ subsets of vertices. Even so, the finite problem is NP-complete, an analog of undecidability, and on infinite graphs it is very likely undecidable. Certainly there is no apparent solution.

The primary motivation for considering the EIP on infinite graphs is to develop global methods. Problems are the life blood of mathematics and there are some very large, i.e. finite but for all practical purposes infinite, graphs for which we would like to solve the EIP. The 120-cell, an exceptional regular solid in four dimensions, is the only regular solid for which we have not solved the EIP. It has 600 vertices so we prefer to call it the 600-vertex, V_{600}. Another is the graph of the n-permutohedron, $n \geq 4$, which has $n!$ vertices. Solving those problems will require developing better methods than we have now. The regular tessellations of Euclidean space are relatively easy to work with but present some of the same kinds of technical problems as those higher dimensional semiregular and exceptional regular solids.

There are also problems arising in applications which bring us to consider isoperimetric problems on infinite graphs. The original application, solving a kind of layout problem if G is regarded as representing an electronic circuit, did not seem to make sense if G is infinite. However we now have a way to make sense of it: Steiglitz and Bernstein [**87**] (see Exercise 1.10 of Chapter 1) noted that in laying out Q_d on a linear chassis, the original problem, which was to minimize the total length of the wires necessary to make the connections, could be generalized to arbitrary spacings between sites, $x_1 < x_2 < ... < x_n$. The same holds for any graph, G, and then the wirelength for a layout function $\varphi : V \to \{1, 2, ..., n\}$, assigning v to $x_{\varphi(v)}$, would just be

$$wl(\varphi) = \sum_{k=0}^{n} (x_{k+1} - x_k) \left| \Theta(S_k(\varphi)) \right|,$$

where $S_k(\varphi) = \{v \in V : \varphi(v) \leq k\}$. Recall that $|S_k(\varphi)| = k$ and $S_k(\varphi) \subset S_{k+1}(\varphi)$. Conversely, if the EIP on G has a nested family of solutions, one for each value of k between 0 and n, which it does for Q_d and many other interesting graphs, then the corresponding layout function is optimal for any choice of the sites, $\{x_k : k = 1, ..., n\}$, even if $n = \infty$. There is then a possibility that the wirelength could be finite and calculable if $\{x_k\}$ is bounded. If G does have nested solutions for the EIP, the finiteness of its wirelength would just depend on the rate of growth of $\min_{|S|=k} |\Theta(S)|$ as $k \to \infty$ and the rate at which $x_{k+1} - x_k \to 0$.

So the serpent of infinity rears its beautiful and awful head in combinatorial paradise, bringing an end to innocence, a beginning of knowledge.

7.1 Euclidean tessellations

A tessellation of $\mathbb{R}^d$ (Coxeter [**28**] calls them honeycombs) can be thought of as a large $(d+1)$-dimensional regular solid in that its building blocks are d-dimensional regular solids. This leads to their Schläfli symbols having $d+1$ entries. Also, their symmetry groups, although infinite and composed of affine orthogonal transformations (i.e. translations as well as reflections and rotations), are still discrete and generated by reflections. They are thus Coxeter groups, treatable by essentially the same theory as the finite Coxeter groups, but having $d+1$ generators.

7.1.1 Cubical

There is one family of tessellations which occur in all dimensions. Its tiles, or cells, are cubes and its vertices may be represented as d-tuples of integers, so $V = \mathbb{Z}^d$. The minimum distance between any two such points is 1, achieved when two d-tuples are identical except in one coordinate where they differ by exactly 1. Edges connect just such pairs. Each vertex, $v \in \mathbb{Z}^d$, has $2d$ neighbors, $v \pm \delta^{(i)}$, $i = 1, ..., d$, so its vertex-figure (see [**28**]) is $\maltese_d$. It is self-dual and its Schläfli symbol is $\left\{4, 3^{d-1}, 4\right\}$. The graph of the cubical tessellation is ubiquitous in mathematics, though often known by other names such as "grid" or "lattice."

7.1.1.1 $d = 1$

The solution of the EIP in the one-dimensional case is easy to guess and prove. The same arguments which work for $\mathbb{Z}_n$, those of Section 1.2.2 of Chapter 1 or Example 3.2.7.2.1 of Chapter 3, also work for $\mathbb{Z}$. It has nested solutions and the initial segments of the numbering $\varepsilon : \mathbb{Z} \to \mathbb{Z}_+$ defined by

$$\varepsilon(i) = \begin{cases} -2i & \text{if } i < 0, \\ 2i + 1 & \text{if } i \geq 0. \end{cases}$$

are optimal.

7.1.1.2 $d = 2$

The two-dimensional case is not trivial but neither is it difficult. For a compressed set, S, $\Theta(S) = 2(I + J)$, where $I = \max\{i_1 - i_2 : (i_1, j_1), (i_2, j_2) \in S\}$ and $J = \max\{j_1 - j_2 : (i_1, j_1), (i_2, j_2) \in S\}$. Also $|S| \leq IJ$ so for $|S| = k = l^2$, a perfect square, the result follows from the classical isoperimetric inequality: of all rectangles of a given boundary, the square has the greatest area. For $l^2 < k < (l+1)^2$ some additional argument is required.

Exercise 7.1 *Complete the argument for $d = 2$.*

Exercise 7.2 *Describe an optimal order for $\mathbb{Z}^2$, consistent with that for $\mathbb{Z}$.*

7.1.1.3 $d > 2$

From the solutions in the cases $d = 1, 2$, one can guess that for $d > 2$ the EIP on $\mathbb{Z}^d$ has nested solutions given by the following recursively defined total order. For $x \in \mathbb{Z}^d$ let

$$\|x\|_\infty = \max\{|x_i| : 1 \leq i \leq d\}.$$

Also for $r \in \mathbb{Z}_+ = \{0, 1, 2, ...\}$ let

$$F_{r,i_0} = \left\{x \in \mathbb{Z}^d : \max x_i = r, \forall i,\ x_i > -r \text{ and } i_0 = \max\{i : x_i = r\}\right\}$$

and

$$F_{-r,i_0} = \left\{x \in \mathbb{Z}^d : \min x_i = -r, \forall i,\ x_i \leq r \text{ and } i_0 = \min\{i : x_i = -r\}\right\}.$$

Note that $F_{0,i} = \left\{0^d\right\}, \forall r > 0, \forall i \neq j, F_{r,i} \cap F_{r,j} = \emptyset$ and

$$\bigcup_{i=1}^{d} F_{r,i} = \left\{x \in \mathbb{Z}^d : \|x\|_\infty = r\right\}.$$

Then for $x, y \in \mathbb{Z}^d$ we define $x <_\varepsilon y$ iff

(1) $\|x\|_\infty < \|y\|_\infty$, or
(2) $\|x\|_\infty = \|y\|_\infty = r > 0$ and
 (a) $x \in F_{r,i_0}$, $y \in F_{r,j_0}$ with $i_0 < j_0$, or
 (b) $x \in F_{r,i_0}$, $y \in F_{-r,j_0}$, or
 (c) $x \in F_{-r,i_0}$, $y \in F_{-r,j_0}$ with $i_0 > j_0$, or
(3) $x, y \in F_{\pm r,i_0}$ and $x' <_\varepsilon y'$, where x' and y' are obtained from x and y, respectively, by removing their i_0th entry ($x_{i_0} = y_{i_0} = \pm r$).

Note that the $F_{\pm r,i}$, $r > 0$, correspond to faces of the d-cube and therefore to vertices of $\maltese_d$. The order of $\left\{F_{\pm r,i}\right\}$ in part (2) of the above definition is that of the (unique) stable numbering of $V_{\maltese_d}$ (see Example 3.2.7.2.2 of Chapter 3). We shall denote the numbering determined by $<_\varepsilon$ as $\varepsilon : \mathbb{Z}^d \to \mathbb{Z}_+$.

Exercise 7.3 *What are the minimal and maximal members of*

$$\left\{x \in \mathbb{Z}^d : \|x\|_\infty = r\right\}$$

wrt ε?

Theorem 7.1 $\forall d \geq 0$, *the EIP on* $\mathbb{Z}^d$ *has nested solutions given by* $\varepsilon : \mathbb{Z}^d \to \mathbb{Z}_+$.

Lemma 7.1 $\varepsilon : \mathbb{Z}^d \to \mathbb{Z}_+$ *is consistent.*

Proof Suppose $x, y \in \mathbb{Z}^d$ and for some i, $1 \leq i \leq d$, $x_i = y_i$. Let x'', y'' be obtained from x and y, respectively, by removing their ith entry. We must show then that $x <_\varepsilon y$ implies $x'' <_\varepsilon y''$.

(1) If $\|x\|_\infty < \|y\|_\infty$ then $\left\|x''\right\|_\infty < \left\|y''\right\|_\infty$ and $x'' <_\varepsilon y''$.
(2) If $x \in F_{r,i_0}$, $y \in F_{r,j_0}$ with $i_0 < j_0$ then $i \neq j_0$ by the definition of i. If $i = i_0$ then either $\left\|x''\right\|_\infty < r$ or $\left\|x''\right\|_\infty = r$ and $x'' \in F_{r,k}$ with $k < i_0$. In either case $x'' <_\varepsilon y''$. If $i \neq i_0$, then $x'' \in F_{r,i_0}$, $y'' \in F_{r,j_0}$ and again, $x'' <_\varepsilon y''$. The subcases (2.b,c) are similar to the first (2.a).
(3) If $x, y \in F_{\pm r,i_0}$ and $i = i_0$, then $x'' = x' <_\varepsilon y' = y''$ so $x'' <_\varepsilon y''$. If $i \neq i_0$ then $x'', y'' \in F_{\pm r,i_0}$ and $x'' <_\varepsilon y''$ iff $\left(x''\right)' <_\varepsilon \left(y''\right)'$. But $x <_\varepsilon y$ implies that $x' <_\varepsilon y'$ so by induction on d, $\left(x'\right)'' <_\varepsilon \left(y'\right)''$. We are then done since $\left(x''\right)' = \left(x'\right)''$.

□

Proof (of Theorem 7.1). We have already proven the theorem for $d = 1, 2$. Assume it true for $d - 1 \geq 2$. Since ε is consistent, compression applies, so if the theorem is not true for d then there exists a compressed set, $S \subseteq \mathbb{Z}^d$, $|S| = k$, and $|E(S)| > |E(S_k(\varepsilon))|$. Assume that S is such a set for which $\varepsilon(S) = \sum_{x \in S} \varepsilon(S)$ is a minimum. Let a be the member of S which maximizes $\varepsilon(x)$ and b be the member of $\mathbb{Z}^d - S$ which minimizes $\varepsilon(x)$. Since $S \neq S_k(\varepsilon)$, $b <_\varepsilon a$ and we consider two cases:

(1) $\|b\|_\infty = \|a\|_\infty$: The $F_{\pm r,i}$, which partition $\{x \in \mathbb{Z}^d : \|x\|_\infty = r\}$, are each totally ordered by the compressibility order, $\mathcal{C}$, since for $x \in F_{\pm r,i}$, $x_i = \pm r$. The minimal member of $F_{r,i}$ is $0^{i-1}r0^{d-i}$ and its maximal member is $r^i(-(r-1))^{d-i}$. Also, the minimal member of $F_{-r,i}$ is $0^{i-1}(-r)0^{d-i}$ and its maximal member is $r^{i-1}(-r)^{d-i+1}$. If $1 < i < d$, the least upper bound (wrt $\mathcal{C}$) of the maximal member of $F_{r,i}$ and the minimal member of $F_{r,i+1}$ is $r0^{i-1}r0^{d-i-1}$ and their greatest lower bound is $0r^{i-1}(-(r-1))^{d-i}$. If $i = 1$ it is $r^2 0^{d-2}$ and $r(-(r-1))0(-(r-1))^{d-3}$ respectively. Since

$$r^i(-(r-1))^{d-i} <_{\mathcal{C}} r0^{i-1}r0^{d-i-1} <_{\mathcal{C}} 0^{i+1}r0^{d-i-2},$$

we need only consider three subcases, corresponding to the three parts of the second case of the definition of $<_\varepsilon$:

(a) $b \in F_{r,i}$ and $a \in F_{r,i+1}$: If $i > 1$, then

$$b \in \left\{xr^{i-1}(-(r-1))^{d-i} : x \neq 0\right\}$$

since

$$0r^{i-1}(-(r-1))^{d-i} <_{\mathcal{C}} 0^i r0^{d-i-1} <_{\mathcal{C}} a.$$

If $r = 1$, our problem reduces to the EIP on Q_d, which we have already solved. Since ε on $\bigcup_{i=1}^d F_{1,i} = \{0,1\}^d$ is essentially lex order, we are done. If $r > 1$, then $\Delta(b) = d \geq \Delta(a)$ and $|S + \{a\} - \{b\}| = |S|$, $|E(S + \{a\} - \{b\})| \geq |E(S)|$ and $\varepsilon(S + \{a\} - \{b\}) < \varepsilon(S)$, a contradiction. If $i = 1$, then

$$b \in \left\{rx(-(r-1))^{d-2}\right\} \cup \left\{r(-(r-1))x(-(r-1))^{d-3} : x \neq 0\right\}$$

since

$$r(-(r-1))0(-(r-1))^{d-3} <_{\mathcal{C}} 0r0^{d-2} <_{\mathcal{C}} a.$$

If $b \neq r0(-(r-1))^{d-2}$, then $S + \{a\} - \{b\}$ leads to the same contradiction. If $b = r0(-(r-1))^{d-2}$ (so $\Delta(b) = d - 1 < d$), then we note that

$$a \in \left\{xr0^{d-2} : -r < x < r\right\}$$

since

$$b <_{\mathcal{C}} r(-(r-1))^{d-1} <_{\mathcal{C}} r^2 0^{d-2}.$$

So if we let

$$A = \left\{ xr0^{d-2} : \varepsilon(x) \leq \varepsilon(a_1) \right\}$$

and

$$B = \left\{ rx\left(-(r-1)\right)^{d-2} : \varepsilon(x) \leq \varepsilon(a_1) \right\},$$

then

$$|S + A - B| = |S|,$$
$$|E(S + A - B)| = |E(S)| \text{ and}$$
$$\varepsilon(S + A - B) < \varepsilon(S),$$

a contradiction.

(b) $b \in F_{r,d}$ and $a \in F_{-r,d}$

(c) $b \in F_{-r,d}$ and $a \in F_{-(r-1),d}$

(2) $\|b\|_\infty < \|a\|_\infty$

Exercise 7.4 *Complete the proof of any one of the cases (1b), (1c) or (2).*

□

7.1.2 Triangular

Aside from the cubical tessellations, with one of each dimension so they form an infinite family, there are only finitely many other tessellations of Euclidean space (see [**28**]), all of dimension 2 or 3. The tessellation of the Euclidean plane by equilateral triangles is a familiar one (Fig. 7.1). Its Schläfli symbol is $\{3, 6\}$ [**28**], reflecting the fact that every face (tile) is bounded by three edges and every vertex is incident to six edges.

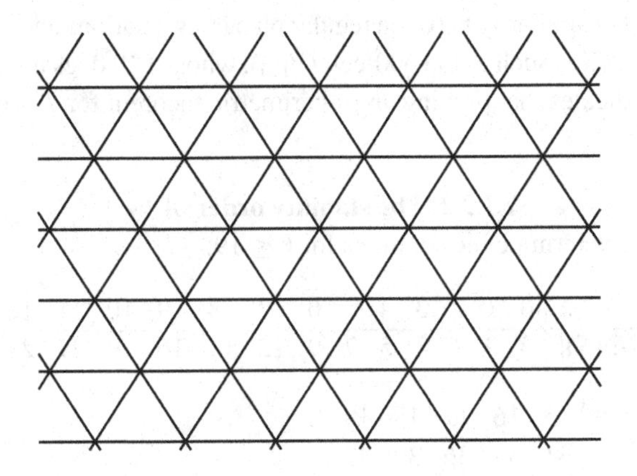

Fig. 7.1 The triangular tessellation, T.

The EIP on the graph, T, of this tessellation is qualitatively different from that of the square tessellation, $\mathbb{Z}^2$, since T is not a product and compression does not apply.

7.1.2.1 What are the solutions?

In order to gain some idea of what the solution sets for the edge-isoperimetric problem on the graph, T, are, we begin in the usual fashion, determining them for small values of k.

$k = 0$: This is trivial since only the null set, $\emptyset$, is of size 0, so it is the unique solution and $\min_{|S|=0} |\Theta(S)| = 0$.

$k = 1$: There are countably many 1-sets of vertices but they are all equivalent under symmetry so they are all solutions and $\min_{|S|-1} |\Theta(S)| = 6$.

$k = 2$: There are countably many equivalence classes of 2-sets under symmetries of the triangular tessellation. The problem is solvable though, since every pair of vertices is either connected by an edge or not. If they are, $|\Theta(S)| = 10$, otherwise $|\Theta(S)| = 12$. Therefore $\min_{|S|=2} |\Theta(S)| = 10$.

$k = 3$: There is only one type of set with $|S| = 3$ and $|E(S)| \geq 3$, the vertices of a triangle. Therefore $\min_{|S|=3} |\Theta(S)| = 6 \cdot 3 - 2 \cdot 3 = 12$ by the remarks of Section 1.2.2 of Chapter 1.

The challenge of the problem for $k > 3$ is apparent. There are countably many equivalence classes of k-sets, of increasing complexity. Even if we could characterize them all, we would still need something stronger than symmetry, something which would systematically take the connectivity of k-sets into account. Fortunately we have just such a tool available, the theory of stabilization (see Chapters 3 and 5) which utilizes Coxeter theory. It is not difficult to come up with a persuasive conjecture about the solution of the EIP on T, but proving it is a different matter. Isoperimetric theorems are notoriously slippery to prove anyway and the similarity between regular planar tessellations and regular four-dimensional solids such as the 600-cell (V_{120}), whose EIP does not have nested solutions, indicates that proving an isoperimetric theorem for T requires some subtlety.

7.1.2.2 The stability order of V_T

From this we determine the solutions for $k \leq 19$:

k	0	1	2	3	4	5	6	7	8	9	10	11	12	13	14
$\max_{\lvert S\rvert=k} \lvert E(S)\rvert$	0	0	1	3	5	7	9	12	14	16	19	21	24	26	29

15	16	17	18	19
31	34	36	39	42

7.1.2.3 Solutions for all k

Theorem 7.2 *T has nested solutions for the edge-isoperimetric problem, i.e. there exists a total order τ on V_T such that for all $k \in \mathbb{Z}^+$ the initial k-set of τ maximizes $|E(S)|$ over all $S \subseteq V_T$ with $|S| = k$.*

For any $v, w \in V_T$ let $d(v, w)$ denote the minimum length of any path from v to w in T and for any $r \in \mathbb{Z}^+$ let

$$B_r = \{v \in V_T : d(v_0, v) \leq r\},$$

the ball of radius r centered at v_0, v_0 being the unique vertex of T which is in the fundamental chamber (labeled a in Fig. 7.2). The sides of B_r for $r > 0$ lie in six straight lines, i.e. B_r has the shape of a regular hexagon. From this it is easy to see that $|B_r| = 1 + 3r(r + 1)$.

All the edges of T lie on three families of parallel lines, which we denote altogether as $\mathcal{L}$, and each vertex lies at the intersection of three of these lines, one from each family. V_3, the group of symmetries of T (see Table IV of [**28**]), acts transitively on $\mathcal{L}$. The theory of stabilization developed in Chapters 3 and 5 is for vertices (points in $\mathbb{R}^d$) but applies equally well to geometric objects, such as lines, which form a closed set, such as $\mathcal{L}$, under the action of a Coxeter group. If $\mathcal{R}$ is a reflection in V_3 we let

$$\mathcal{S}(\mathcal{L}; \mathcal{R}; p) = \{(L, \mathcal{R}(L)) : \|L - p\| < \|\mathcal{R}(L) - p\|\}.$$

Then the *stability order of $\mathcal{L}$ with respect to V_3 and p*, denoted $\mathcal{S}(\mathcal{L}; V_3; p)$, is the transitive closure of

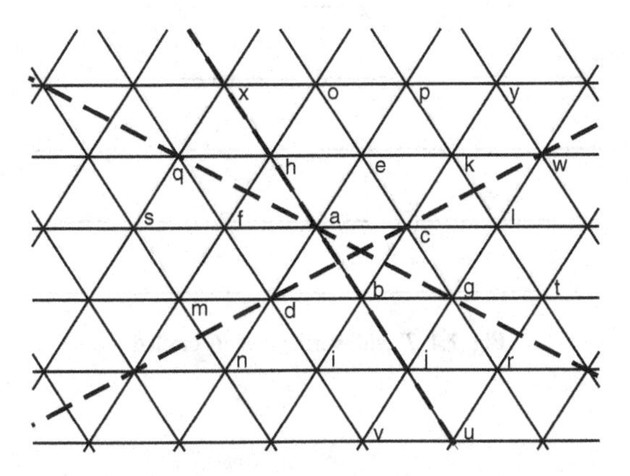

Fig. 7.2 T with the fixed lines of basic reflections dashed.

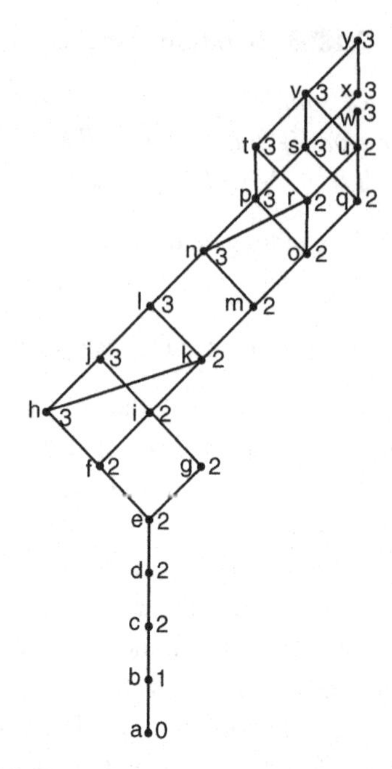

Fig. 7.3 The stability order, $\mathcal{S}(T)$, with weight, Δ.

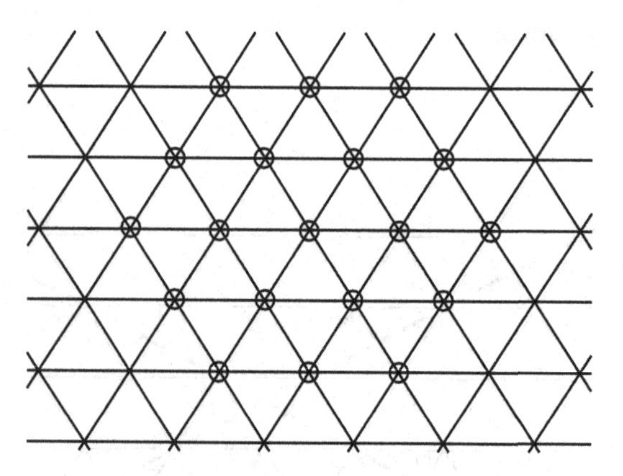

Fig. 7.4 T with vertices of B_2 circled.

$$\bigcup_{\mathcal{R} \in V_3} \mathcal{S}(\mathcal{L}; \mathcal{R}; p).$$

Compare this to Definition 3.2.7.1 of Chapter 3. The symmetries of B_r constitute a dihedral group, $I_2(6)$, a subgroup of V_3, which act transitively on the six lines bounding B_r. The stability order of any $I_2(n)$ acting on the sides of a regular n-gon is total (see Example 1 of Section 3.2.7 of Chapter 3) so the relative order (in $\mathcal{S}(\mathcal{L}; V_3; p)$) of the six lines bounding B_r is total. Thus we may denote them as $L_{r,i}$, $0 \leq i \leq 5$ with $L_{r,i} <_{\mathcal{S}} L_{r,j}$ if $i < j$.

Lemma 7.2 *Lines $L_{r,i}$ and $L_{s,j}$ are incomparable in $\mathcal{S}(\mathcal{L}; V_3; p)$ iffs $= r + 1$, $i = 5$ and $j = 0$.*

Proof $L_{r,5}$ intersects $L_{r+1,1}$ and the lines which bisect the angles between them are lines of symmetry for T, so $L_{r,5} <_{\mathcal{S}} L_{r+1,1}$. Similarly, $L_{r,4} <_{\mathcal{S}} L_{r+1,0}$. However, $L_{r,5}$ and $L_{r+1,0}$ are parallel and the bisector of any perpendicular which connects them is not a line of symmetry for T. If we could show that $L_{r,5} <_{\mathcal{S}} L_{r+1,0}$ it would have to be because there is a line L such that $L_{r,5} <_{\mathcal{S}} L <_{\mathcal{S}} L_{r+1,0}$ but $\forall L \in \mathcal{L} - \{L_{r,5}, L_{r+1,0}\}$, $L <_{\mathcal{S}} L_{r,5}$ or $L_{r+1,0} <_{\mathcal{S}} L$, so we are done. □

We have

$$V_T = \{v_0\} \cup \bigcup_{r=1}^{\infty} \bigcup_{i=0}^{5} (V_T \cap L_{r,i})$$

Each vertex, except v_0, is contained in multiple $L_{r,i}$'s but if we let

$$L'_{r,i} = V_T \cap L_{r,i} - \left[\bigcup_{j=i+1}^{5} (V_T \cap L_{r,j}) \cup \bigcup_{s=r+1}^{\infty} \bigcup_{i=0}^{5} (V_T \cap L_{s,i}) \right]$$

then $\{v_0\}$ and the $L'_{r,i}$ partition V_T. Also, $B_r = \{v_0\} \cup \bigcup_{s=1}^{r} \bigcup_{i=0}^{5} L'_{s,i}$ and

$$\left| L'_{r,i} \right| = \begin{cases} r-1 & \text{if } i = 0, \\ r & \text{if } 0 < i < 5, \\ r+1 & \text{if } i = 5. \end{cases}$$

Note that $V_T \cap L_{r,i}$ is totally ordered by $\mathcal{S}(\mathcal{L}; V_3; p)$, the vertex nearest p being its least element of course and this lies at or near the midpoint of $B_r \cap L_{r,i}$. The others follow in increasing order of their distance from p so that they alternate from side to side. $L'_{r,i}$ is an initial segment in this order. Note also that the perpendicular bisector of every edge of T is the fixed line of a reflective symmetry. The ends of the edge are therefore comparable in its stability order and for every $v \in V_T$ we may define a weight

$$\Delta(v) = |\{w \in V_T : \exists v \in E_T,\ \partial(e) = \{v, w\}\ \&\ w <_S v\}|$$
$$= \begin{cases} 0 & \text{if } v = v_0 \in B_0, \\ 1 & \text{if } v = v_1 \in L'_{1,1}, \\ 2 & \text{if } v \neq v_0, v_1 \text{ \& minimal in } L'_{r,i}, \\ 3 & \text{otherwise.} \end{cases}$$

and then $\forall S \subseteq V_T$,

$$|E(S)| = \sum_{v \in S} \Delta(v).$$

Proof (of Theorem 7.2). We define a total order, τ, on V_T by $v <_\tau w$ if $v \in L'_{r,i}$, $w \in L'_{s,j}$ and

(1) $r < s$, or
(2) $r = s$ & $i < j$, or
(3) $r = s$ & $i = j$ & $v <_S w$.

Note that $v <_S w$ implies $v <_\tau w$, i.e. $<_\tau$ is an extension of $\mathcal{S}(\mathcal{L}; V_3; p)$. Let $\tau : V_T \to \mathbb{Z}^+$ denote the numbering determined by $<_\tau$. By the theory of stabilization, we need only show that if $S \subseteq V_T$ is stable, i.e. a lower set in the stability order, $\mathcal{S}(\mathcal{L}; V_3; p)$, $|S| = k$ then

$$|E(S_k(\tau))| \geq |E(S)|.$$

If $S \neq S_k(\tau)$, then $\exists$ a minimal element, a, with respect to $<_\tau$, in $V_T - S$ and a maximal element, b, in S. Note that $a <_\tau b$ but they must be incomparable with respect to $\mathcal{S}(\mathcal{L}; V_3; p)$. Having already proved the theorem for $k = 0, 1, 2$ we may assume $k > 2$ so $\Delta(a), \Delta(b) = 2$ or 3. If $\Delta(a) \geq \Delta(b)$ then $|S + \{a\} - \{b\}| = k$ and $|E(S + \{a\} - \{b\})| \geq |E(S)|$ and a finite series of such switches will achieve our goal. The only way $\Delta(a) < \Delta(b)$ is if a is the minimal element of $L'_{r,5}$, for some r, and $b \in L'_{r+1,0}$ but not minimal. Then switching all of $S \cap L'_{r+1,0}$ for the initial segment of $L'_{r,5}$ of the same size will do the job. This is possible because $\left|L'_{r+1,0}\right| = (r+1) - 1 = r < r + 1 = \left|L'_{r,5}\right|$. □

Corollary 7.1 *If $k = 1 + 3r(r+1)$ then the only stable solution is B_r.*

Corollary 7.2 *$<_\tau$ is the only total extension of $\mathcal{S}(\mathcal{L}; V_3; p)$ whose initial segments are solutions of the EIP.*

There is another, in some ways more natural, total ordering, τ' of V_T whose initial segments are solutions of the EIP: Begin with $v_0 <_{\tau'} v_1$ and having chosen $v_0 <_{\tau'} v_1 <_{\tau'} ... <_{\tau'} v_n$ choose v_{n+1} to be the furthest counterclockwise neighbor of v_n which has not been chosen yet (see Fig. 7.5). The initial segments of this total order are not stable but the sequence of marginal contributions of the vertices is the same as the sequence of weights with respect to τ. The clockwise spiral works equally well, of course.

7.1.3 The hexagonal tessellation

The tessellation of the Euclidean plane by regular hexagons is also familiar from beeswax and bathroom floors (see Fig. 7.6). It is the dual of the triangular tessellation and has Schläfli symbol $\{6, 3\}$ (see [**28**]) meaning that every face (tile) is bounded by six edges and every vertex is incident to three edges. Let H denote the graph of this tessellation.

The steps which we took to find and prove the solution for T also suffice for H. There are some minor complications since the solution sets are no longer

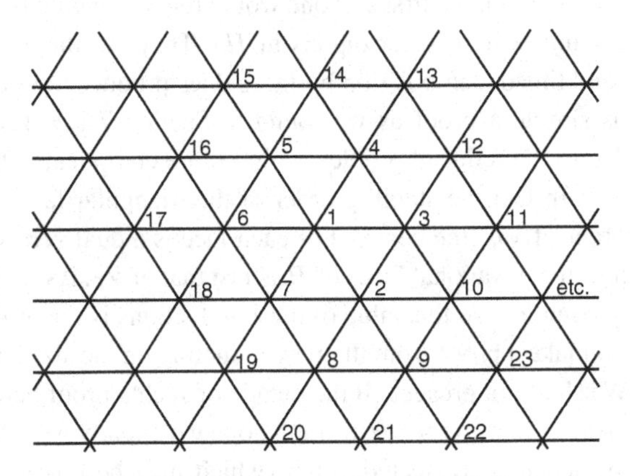

Fig. 7.5 T with an optimal spiral numbering.

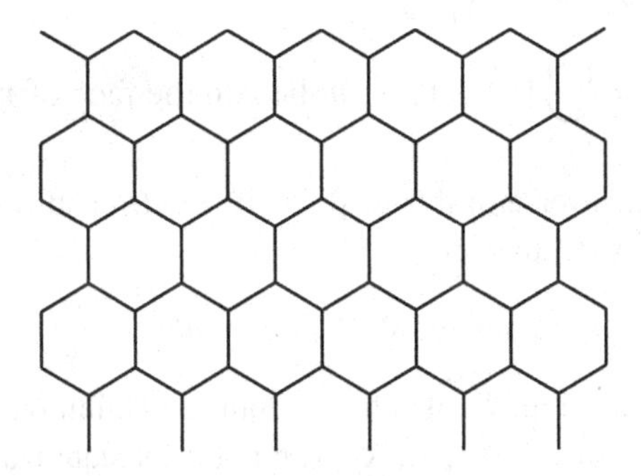

Fig. 7.6 The hexagonal tessellation, H.

balls in an intrinsic metric and their boundary vertices do not lie on straight lines but zig-zag a bit; however, the same program does work. For later reference, we list the solutions for $k \leq 24$:

k	0	1	2	3	4	5	6	7	8	9	10	11	12	13	14
$\max_{\|S\|=k} \|E(S)\|$	0	0	1	2	3	4	6	7	8	9	11	12	13	15	16

k	15	16	17	18	19	20	21	22	23	24
$\max_{\|S\|=k} \|E(S)\|$	17	19	20	21	23	24	25	27	28	30

Exercise 7.5 *Use stabilization to verify the values of* $\max_{|S|=k} |E(S)|$ *for* H *with* $0 \leq k \leq 24$.

However, since our purpose here is to develop methods as well as to solve problems, we shall proceed a little differently. In the proof of Theorem 7.2 we decomposed V_T into "lines". In order for such a decomposition to work, the blocks of the partition (the $L'_{r,i}$ in that case) must be highly connected. With that requirement in mind, the first sets one would think of would be the vertices of faces (i.e. triangles in T or hexagons in H). They are the sets of highest connectivity, and this decomposition by faces does produce a proof for T and H, just not as simple a proof as we found in Section 7.1.2. The additional difficulty is due to the reduced problem being more complicated than the one in Section 7.1.2. In fact the stability order of the triangular faces in T is just the stability order of V_H (the centroid of each face is a dual vertex). Also, the stability order of the hexagonal faces of H is just that of V_T. As we know, these are both fairly complex, so the value of these reductions is not at first evident. It is possible to make a proof from them by inducting on the sizes of subsets of V_H and V_T. We shall not present all the details of such a proof here, just those necessary to get the solution for H from the one we already have for T.

If G is any planar 3-connected graph (which may be represented on the surface of a sphere if finite) let G^* be its dual. Then $H^* = T$, $T^* = H$ and $G^{**} = G$ in general. If $S \subseteq V_G$, let

$$S^* = \bigcup_{v \in S} \{w \in V_{G^*} : w \text{ lies on the face of } v\}.$$

Also, let $S^{\underline{*}}$ be the inverse of this map, i.e. if $S = U^*$ and U is maximal with respect to this property, then $S^{\underline{*}} = U$.

Lemma 7.3 *$S \subseteq V_G$ is stable iff $S^* \subseteq V_{G^*}$ is stable.*

Proof For any reflection, $\mathcal{R}$, it follows from the definition of $Stab_{\mathcal{R},p}(S)$. Being true for each of $\mathcal{R}_0, \mathcal{R}_1, ..., \mathcal{R}_{k-1}$, it holds for stabilization with respect to the whole set. □

One would further expect that the optimality of S and S^* would be closely connected but

Exercise 7.6 *Find a set $S \subseteq V_H$ which is optimal but $S^* \subseteq V_T$ is not.*

Exercise 7.7 *Find a set $S \subseteq V_T$ which is not optimal but $S^* \subseteq V_H$ is.*

So the optimality of S and S^* are logically independent.

Definition 7.1 *$k \in \mathbb{Z}^+$ is called a critical cardinal if*

$$\max_{|S|=k} |E(S)| - \max_{|S|=k-1} |E(S)| > 1.$$

Note that if S is optimal and has a pendant vertex, v, then $S - \{v\}$ must also be optimal since $|E(S)| = |E(S - \{v\})| + 1$, the least that $|E|$ can increase. Thus such an S is not critically optimal. Therefore, if S is critically optimal it can have no pendant vertices and must be a union of faces. So for a critically optimal S, $S^{\underline{*}}$ is defined.

Recalling the definition of induced edges in Section 1.2.2.(4) of Chapter 1, we introduce the notion of *incident* edges for $S \subseteq V$:

$$I(S) = \{e \in E : \partial(e) = \{v, w\},\ v \in S \text{ or } w \in S\}.$$

Then the *incident edge problem* on a graph is to minimize $|I(S)|$ over all $S \subseteq V$ with $|S| = k$.

Lemma 7.4 *If $G = (V, E, \partial)$ is a regular graph of degree δ and $S \subseteq V$, then $\forall S \subseteq V$,*

$$I(S) = E(S) + \Theta(S)$$

so

$$\begin{aligned} |I(S)| &= \frac{1}{2}(\delta|S| - |\Theta(S)|) + |\Theta(S)| \\ &= \frac{1}{2}(\delta|S| + |\Theta(S)|). \end{aligned}$$

Corollary 7.3 *If G is a regular graph, then $S \subseteq V$ is a solution of the incident edge problem iff it is a solution of the EIP. Also, $\forall k, \min_{|S|=k} |\Theta(S)| = 2\min_{|S|=k} |I(S)| - \delta k$.*

Lemma 7.5 *If $S \subseteq V_G$ is optimal and $|S| = k$, a critical cardinal, then $S^{\underline{*}} \subseteq V_{G^*}$ is optimal.*

Proof Euler's relation, $v + f = e + 2$, holds for the subgraph of G induced by S with $v = |S|$, $f = |S^{\underline{*}}|$ and $e = |E(S)| = |I(S^{\underline{*}})|$. If $S^{\underline{*}}$ is not optimal then $\exists U \subseteq V_{G^*}$ such that $|U| = |S^{\underline{*}}|$ and $|I(U)| < |I(S^{\underline{*}})|$. The Euler relation then implies that $|U^*| < |S|$. In fact $|S| - |U^*| = |I(S^{\underline{*}})| - |I(U)| = |E(S)| - |E(U^*)|$. Adding $|S| - |U^*|$ vertices to U^* optimally will give a set S' such that $|S'| = |S|$ and $|E(S')| \geq |E(S)|$. This contradicts the optimality of S if $|E(S')| > |E(S)|$ or its criticality if $|E(S')| = |E(S)|$. □

Theorem 7.3 *H has nested solutions for the edge-isoperimetric problem, i.e. there exists a total order on V_H whose initial k-set minimizes $|\Theta(S)|$ over all $S \subseteq V_H$ with $|S| = k$ for all $k \in \mathbb{Z}^+$.*

Proof If k is a critical cardinal for H, and S is a stable optimal set of size k, then by Lemma 7.5 $S^{\underline{*}}$ is optimal in $H^* = T$. By Lemma 7.3 it is also stable in T so must be as described in the proof of Theorem 1.1. For

$$1 + 3r(r+1) \leq k' \leq 1 + 3r(r+1) + r + 4(r+1)$$

this set is uniquely determined and $S = (S^{\underline{*}})^*$ is also. If

$$1 + 3r(r+1) + r + 4(r+1) < k' < 1 + 3(r+1)(r+2)$$

then there are two possibilities but since $|E(U)|$ is the same for both, they are both optimal. In particular, $S^*_{k'}$ is optimal for $k' = 1, 2, \ldots$ For the noncritical cardinals we need only interpolate the vertices in $(S_{k'+1})^* - S^*_{k'}$, in the order determined by stabilization, to prove the theorem. □

Corollary 7.4 *If $k = 6(r+1)^2$ then the only stable solution is B^*_r.*

There are also optimal spiral (clockwise and counterclockwise) orderings of V_H.

7.2 Comments

A solution of the EIP on $\mathbb{Z}^2$ was given by Harary and Harborth [**44**]. The solution of the EIP on $\mathbb{Z}^d$ was first carried out by Bollobas and Leader [**22**] (using compression) and later by Ahlswede and Bezrukov [**1**]. Bollobas and Leader actually proved a stronger result, solving the induced EIP for finite grids, P_n^m. The result for grids is one of the few for graphs which are not regular, but this may be explained by the fact that the induced EIP on P_n^m is essentially equivalent to that for $\mathbb{Z}^m$ which is regular. Ahlswede and Bezrukov extended the Bollobas–Leader results to products of distinct factors, $P_{n_1} \times P_{n_2} \times \ldots \times P_{n_m}$.

It is at first surprising, but easy to verify, that the results for finite graphs imply those for the infinite.

Up to now, compression has been the dominant tool for solving isoperimetric problems, but one can now ask if it is possible to give a proof of the EIP for $\mathbb{Z}^d$ or Q_d with stabilization and the other methods developed in this chapter for the exceptional regular tesselations, T and H. We shall return to this question in Chapter 9.

7.2.1 The last Euclidean tessellations

The EIP has not yet been solved for the dual pair of exceptional tessellations of $\mathbb{R}^4$, whose cells are ✠$_4$ and V_{24} respectively (see [**28**], Table II on p. 296).

7.2.2 Powers of T and H

$\mathbb{Z}$ is a tessellation of $\mathbb{R}$ and, as we showed in Section 7.1.1, the EIP on $\mathbb{Z}^d$ has nested solutions for all d. What about T^d for $d \geq 2$; does it have nested solutions?

Exercise 7.8 *Show that T^2 does not have nested solutions for EIP. (Hint: assuming nested solutions, find an optimal compressed set for $k = 14$ and show that it is not optimal.)*

Exercise 7.9 *Show that H^2 does not have nested solutions for EIP. (Hint: same as the preceding except that $k = 12$.)*

7.2.3 Tessellations of hyperbolic space

The connection between Euclidean geometry and combinatorics which seemed implicit in the theory of stabilization was puzzling from its inception. It now appears possible to penetrate that mystery a bit. From hyperbolic geometry we learn that the hyperbolic plane, H_2, also has regular tessellations. The symmetry groups of these tessellations are, in an abstract sense, Coxeter, and correspond to solutions of the inequality

$$\frac{1}{p} + \frac{1}{q} < \frac{1}{2}, \quad p, q \in \mathbb{Z}^+.$$

They occur in dual pairs whose Schläfli symbols are $\{p, q\}$ and $\{q, p\}$. There are infinitely many such, in contrast to the Euclidean condition

$$\frac{1}{p} + \frac{1}{q} = \frac{1}{2}, \quad p, q \in \mathbb{Z}^+,$$

which only has the solutions $\frac{1}{4} + \frac{1}{4} = \frac{1}{3} + \frac{1}{6} = \frac{1}{6} + \frac{1}{3} = \frac{1}{2}$ giving the three tessellations which we have already treated. We conjecture that the methods of this chapter will show that all of the regular tessellations of the hyperbolic plane have nested solutions for the EIP.

There are also exceptional regular tessellations of H_3 with dodecahedral cells which remain to be investigated.

7.2.4 The VIP on $\mathbb{Z}^d$

D.-L Wang and P. Wang [**92**], motivated by problems in theoretical computer science, solved the VIP on $\mathbb{Z}^d$ already in 1977. This is one of the few cases in which the vertex version of an isoperimetric problem was solved before the edge version. Also, this result on infinite graphs preceded the finite one (the Chvátalová–Moghadam theorem for the VIP on $\mathbb{Z}_n^d$ which was mentioned in the Comments section of Chapter 6). However, that reversal of order is not unprecedented. Macaulay's theorem (see Chapter 8), published in 1927 and to which the Wang–Wang theorem is essentially equivalent, is generally recognized as the first solution of a combinatorial isoperimetric problem. Some might even argue that infinite combinatorial isoperimetric problems are more natural than the finite. Wang and Wang used stabilization to prove their result and the argument can be cleaned up along the lines of the proof of Theorem 7.1.

8
Isoperimetric problems on complexes

We find in the literature three distinct connections between combinatorial isoperimetric problems and partially ordered sets:

(1) The reduction of edge and vertex-isoperimetric problems on graphs (EIP and VIP) to maximum weight ideal (MWI) problems on the compressibility or stability order (see Chapters 3, 4, 5 and 6).

(2) J. B. Kruskal's observation, in 1969 [**66**], that a graph may be looked upon as a two-dimensional complex and then its (incident) EIP has a natural extension to arbitrary complexes (hypergraphs). The extension is called the *minimum shadow problem* (MSP). Kruskal had already solved the MSP [**65**] for the simplex in all dimensions, a result discovered independently by G. O. H. Katona [**58**]. The Kruskal–Katona theorem is probably the most widely known and applied of all combinatorial isoperimetric theorems. Kruskal went on in [**66**] to conjecture that our solution of the EIP on Q_d, the graph of the d-cube (see Chapters 1 and 3), could be extended to the MSP on the complex of faces of the d-cube. He also suggested looking for more such analogs of the Kruskal–Katona theorem.

(3) Scheduling problems are standard fare in applied combinatorial optimization. If the steps of a manufacturing process must be carried out in some serial order subject to given precedence constraints, and we wish to order the steps so as to minimize some functional of the ordering, such as the average time between when a step is completed and its last successor is completed, then we have a scheduling problem. Scheduling problems are closely related to layout problems, but instead of an undirected graph, representing a wiring diagram, we have an acyclic directed graph, the Hasse diagram of the poset given by the precedence constraints.

In this chapter we explore these synergistic interactions between isoperimetric problems and partial orders, and extend the global theory of isoperimetric problems on graphs to minimum shadow and scheduling problems.

8.1 Minimum shadow problems

8.1.1 Combinatorial complexes

$\mathcal{F}(P)$, the set of all faces of a d-dimensional convex polytope P (see p. 57 of [**93**]), partially ordered by inclusion, has many special properties but we shall focus on the fact that $\mathcal{F}(P)$ is a lattice with rank function $r(f) = 1 + \dim(f)$, and that it has a minimum element, $\mathbf{0}$ (the null face) and maximum element, $\mathbf{1} = P$. This means that $\mathcal{F}(P)$ is the disjoint union of the sets $\mathcal{F}_r = \{f \in \mathcal{F}(P) : r(f) = r\}$, $0 \leq r \leq d+1$, where $d = \dim(P)$, $\mathcal{F}_0 = \{\mathbf{0}\}$ and $\mathcal{F}_{d+1} = \{P\}$. Convex polytopes whose face lattices are isomorphic are called *combinatorially equivalent* and we shall not distinguish between them. Almost everything we do with face lattices of convex polytopes works equally well for posets with rank function, $\mathbf{0}$ and $\mathbf{1}$, which we denote as *(combinatorial) complexes.* Given a complex, $\mathcal{F}$, a *boundary function*, $\partial_r : \mathcal{F}_r \to 2^{\mathcal{F}_{r-1}}$, $1 \leq r \leq d+1$, is defined by

$$\partial_r(f) = \{g \in \mathcal{F}_{r-1} : g \text{ is contained in } f\}.$$

8.1.1.1 The face lattice of the d-cube

Recall Example 3.2.1.2 of Chapter 3 where the vertices of the d-cube are $V_{\mathbb{Q}_d} = \{-1, 1\}^d$. As a subset of points in $\mathbb{R}^d$, these points span a convex solid, the d-cube itself, whose set of nonempty faces may be represented by

$$\{-1, 1, \omega\}^d$$

with

$$-1 < \omega,$$
$$1 < \omega,$$

and coordinatewise order on the product. The *dimension* of a "face" is the number of ωs it contains; ω may be thought of as a continuous variable taking values between -1 and 1 independently in each coordinate where it appears. Thus ω^d represents the whole d-cube. The face lattice, $\mathcal{Q}_d$, of the d-cube, may

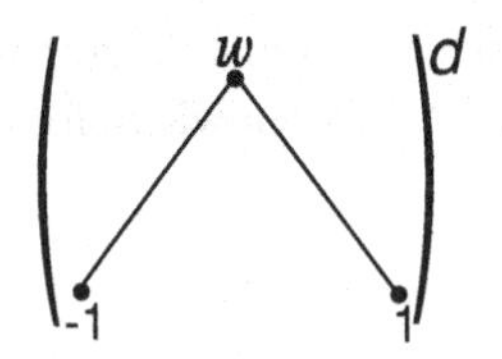

Fig. 8.1.

be represented as the product shown in Fig. 8.1 (see [**21**] for the definition of poset product)

$\partial_r : \mathcal{Q}_{d,r} \to 2^{\mathcal{Q}_{d,r-1}}$ is defined by

$$\partial_r(f) = \left\{ g : \exists j \text{ such that } f_j = \omega, \forall k \neq j, g_k = f_k \right\}.$$

Note that g_j must then be -1 or 1. We may, when convenient, represent the face lattice of the d-cube with -1 replaced by 0.

8.1.1.2 The face lattice of the d-simplex

The vertices of the d-simplex are the $(d+1)$-tuples $\delta^{(i)}$, $1 \leq i \leq d+1$, where

$$\delta_j^{(i)} = \begin{cases} 1 & \text{if } j = i \\ 0 & \text{otherwise.} \end{cases}$$

The face lattice of the d-simplex may then be represented by

$$\{0 < 1\}^{d+1}$$

and the dimension of a "face" is one less than the number of 1s in its representation. Note that our representation of the face lattice of the d-simplex is also the standard representation of the Boolean lattice, $\mathcal{B}_n$, on $n = d+1$ generators.

Exercise 8.1 *Define the boundary function* $\partial_r : \mathcal{B}_{n,r} \to 2^{\mathcal{B}_{n,r-1}}$ *for the simplex.*

8.1.2 Shadows

The boundary function of a face lattice may be extended to subsets of faces, $\partial_r : 2^{F_r} \to 2^{F_{r-1}}$, by

$$\partial_r(S) = \bigcup_{f \in S} \partial_r(f)$$

for $S \subseteq F_r$. $\partial_r(S)$ is also called *the shadow of* S. The *minimum shadow problem* (MSP) on a face lattice $\mathcal{F}$ is, given r and $k \in \mathbb{Z}_+$, to compute

$$\min_{\substack{S \subseteq \mathcal{F}_r \\ |S| = k}} |\partial_r(S)|.$$

We shall see in the following section that on graphs, the MSP is equivalent to the incident edge-isoperimetric problem which for regular graphs is equivalent to the EIP.

8.1.3 Duality

The *dual* (or *opposite*) of a poset $\mathcal{P} = (P, <_{\mathcal{P}})$ is the poset $\mathcal{P}^* = (P, <_{\mathcal{P}^*})$ where $x <_{\mathcal{P}^*} y$ iff $y <_{\mathcal{P}} x$. Note that

$$\mathcal{P}^{**} = \left(\mathcal{P}^*\right)^* = \mathcal{P}.$$

Exercise 8.2 *Duality commutes with product, i.e. for any posets, $\mathcal{P}$, $\mathcal{Q}$,*

$$(\mathcal{P} \times \mathcal{Q})^* \simeq \mathcal{P}^* \times \mathcal{Q}^*.$$

It can be shown (see Theorem 2.7.iv of [**93**]) that the dual of the face lattice of any convex polytope is itself the face lattice of a convex polytope. Informally, the dual of a convex polytope may be thought of as the polytope generated by the centroids of the $(d-1)$-faces of the given d-polytope. This leads to technical difficulties so the rigorous discussion of Ziegler in [**93**] defines the dual (which he calls "the polar") of a convex polytope with the origin in its interior, in terms of linear inequalities.

Example 8.1 $\mathcal{B}_n^* = \mathcal{B}_n$*; the simplex is self-dual.*

Example 8.2 $\mathcal{Q}_d^* = \maltese_d$*; the dual of the d-cube is the d-crosspolytope.*

Exercise 8.3 *Prove that if $\partial_r : \mathcal{F}_r \to 2^{\mathcal{F}_{r-1}}$, $1 \le r \le d$, is the sequence of boundary functions for a face lattice $\mathcal{F}$, then $\partial_r^* : \mathcal{F}_r^* \to 2^{\mathcal{F}_{i-1}^*}$, $1 \le i \le d$, defined by*

$$\partial_i^*\left(f^*\right) = \left\{g^* \in \mathcal{F}_{i-1}^* = \mathcal{F}_{d-i} : f^* \in \partial_{d-i}\left(g^*\right)\right\}.$$

is the sequence of boundary functions for $\mathcal{F}^*$.

Exercise 8.4 *Use the informal definition of the dual to verify that the graph of the dual of the d-simplex is K_{d+1}, the graph of the d-simplex*

Exercise 8.5 *Similarly, verify that the graph of $\mathcal{Q}_d^*$ is $\maltese_d$.*

Exercise 8.6 *Find a Hasse diagram for the face lattice of the crosspolytope, $\maltese_d$.*

Is there any connection between the MSP for $\mathcal{F}$ and $\mathcal{F}^*$? Obviously, for a given r, the function $\min_{\substack{S\subseteq\mathcal{F}_r\\|S|=k}} |\partial_r(S)|$, of k, is nondecreasing. An integer k, $0 \leq k \leq |\mathcal{F}_i|$, is called *critical* if

$$\min_{\substack{S\subseteq\mathcal{F}_r\\|S|=k}} |\partial_r(S)| < \min_{\substack{S\subseteq\mathcal{F}_r\\|S|=k+1}} |\partial_r(S)|.$$

Note that we need only solve the MSP on $\mathcal{F}_r$ for critical cardinals. Solutions for any other cardinal can be obtained by deleting members from an optimal set of the next largest critical size since such deletions cannot decrease the shadow. Then we have

Theorem 8.1 *k is critical for $\mathcal{F}_r$ iff $|\mathcal{F}_{r-1}| - \min_{\substack{S\subseteq\mathcal{F}_i\\|S|=k}} |\partial_r(S)|$ is critical for $\mathcal{F}^*_{d-i}$. In fact, if k is critical for $\mathcal{F}_r$ then $S_0 \subseteq \mathcal{F}_r$, $|S_0| = k$ is optimal iff $T_0 = \mathcal{F}_{r-1} - \partial_r(S_0)$ is optimal for $\mathcal{F}^*_{d-r}$ and $|T_0|$ is critical.*

Proof $|\partial(S_0)| + |T_0| = |\mathcal{F}_{r-1}|$ and $|S_0| + |\partial^*(T_0)| \leq |\mathcal{F}_r|$ (since $S_0 \cap \partial^*(T_0) = \emptyset$). Therefore, if T_0 is not a solution for $\mathcal{F}^*_{d-r}$ then k is not critical for $\mathcal{F}_r$. And if T_0 is a solution but $|T_0|$ is not critical, then S_0 is not a solution. □

Corollary 8.1 *The minimum shadow problems for $\mathcal{F}$ and $\mathcal{F}^*$ are equivalent.*

Now for a graph, $G = (V, E, \partial)$ and $S \subseteq V$, $\partial^*(S)$, the dual shadow of S, is exactly the set of edges incident to S. We noted in Exercise 1.1 of Chapter 1 that in regular graphs, minimizing $|\partial^*(S)|$ over all k-sets of V is equivalent to the EIP. Thus, as we claimed above, the MSP is an extension of the EIP for regular graphs. Furthermore, $\Phi(S) = \partial \circ \partial^*(S) - S$, so the VIP also has a natural extension to complexes.

8.1.4 The Kruskal–Katona theorem

Theorem 8.2 *(Kruskal–Katona) The face lattice of the d-simplex (the Boolean lattice, $\mathcal{B}_{d+1}$) has nested solutions for the minimum shadow problem (MSP). Lexicographic numbering is optimal.*

Proof Recall Theorem 4.2 of Chapter 4, stating that the VIP on Q_n, the graph of the n-cube, has nested solutions given by Hales numbering, H. For any k, $0 \leq k < \binom{n}{r}$, let $k' = \sum_{j=0}^{r-1}\binom{n}{j} + k$ and then $S_{k'}(H)$ minimizes $\Phi(S')$ over all $S' \subseteq V_{Q_n} = \{0,1\}^n$ with $|S'| = k'$. Since $S_{k'}(H)$ is an ideal in $\mathcal{B}_n$, $|S_{k'}(H) \cap \mathcal{B}_{n,r}| =$

k and

$$\begin{aligned}|\Phi(S_{k'}(H))| &= \sum_{j=0}^{r}\binom{n}{j} + |\partial^*(S_k)| - k' \\ &= \binom{d}{r} - k + |\partial^*(S)|,\end{aligned}$$

$S_k = S_{k'}(H) \cap \mathcal{B}_{n,r}$ must minimize $|\partial^*(S)|$ over any $S \subseteq \mathcal{B}_{d,r}$ with $|S| = k$. Thus the terminal segments of $\mathcal{B}_{n,r}$ wrt reverse lex order are solutions of the MSP on $\mathcal{B}_n^*$. Since $\mathcal{B}_n$ is self-dual and the order-reversing isomorphism, complementation of coordinates, reverses lex order and turns terminal segments into initial segments, we are done. □

A couple of months after Kruskal's article [**66**] appeared, G. F. Clements and B. Lindström published [**29**] an extension of the Kruskal–Katona theorem to all products of chains, $\mathcal{T}_{n_1} \times \mathcal{T}_{n_2} \times \ldots \times \mathcal{T}_{n_d}$.

Theorem 8.3 *(Clements–Lindström)* $\mathcal{T}_{n_1} \times \mathcal{T}_{n_2} \times \ldots \times \mathcal{T}_{n_d}$ *has nested solutions for MSP. If* $n_1 \geq n_2 \geq \ldots \geq n_d$*, then lexicographic numbering is optimal.*

The Clements–Lindström theorem may be deduced from the Chvatalova–Moghadam theorem (see Section 5 of Chapter 6), generalizing the above proof of Kruskal–Katona.

8.1.5 Macaulay posets

A complex, $\mathcal{P}$ (ranked poset with **0** and **1**), is called *Macaulay* if $\forall i$, $1 \leq i \leq d$, there exists a numbering $\mu_i : \mathcal{P}_i \to \{1, 2, \ldots, |\mathcal{P}_i|\}$ such that

(1) $\mu_i^{-1}\{1, 2, \ldots, k\}$ is a solution of the MSP, i.e. $\left|\partial_i\left(\mu_i^{-1}\{1, 2, \ldots, k\}\right)\right| = \min_{\substack{S \subseteq \mathcal{P}_i \\ |S|=k}} |\partial_i(S)|$, and

(2) $\partial_i\left(\mu_i^{-1}\{1, 2, \ldots, k\}\right) = \mu_{i-1}^{-1}\left\{1, 2, \ldots, k'\right\}$ where k' is the appropriate minimum value.

Thus the Boolean lattice, $\mathcal{B}_n$, is Macaulay (Theorem 8.2) and more generally, the product of chains $\mathcal{T}_{n_1} \times \mathcal{T}_{n_2} \times \ldots \times \mathcal{T}_{n_d}$ is Macaulay (by Theorem 8.3).

Theorem 8.4 *If* $\mathcal{P}$ *is Macaulay then* $\forall j \geq 0$, $\mu_i^{-1}\{1, 2, \ldots, k\}$ *minimizes*

$$\left|\partial_{i-j} \circ \ldots \circ \partial_{i-1} \circ \partial_i(S)\right|$$

over all $S \subseteq \mathcal{P}_i$ *with* $|S| = k$.

Theorem 8.5 $\mathcal{P}$ *is Macaulay iff* $\mathcal{P}^*$ *is Macaulay.*

Proof It follows from Theorem 8.1 with $\mu_i^*(x) = |\mathcal{P}_{d-i}| + 1 - \mu_{d-i}(x)$. □

If $\mathcal{P}$ is Macaulay, its *rank-greedy Macaulay* numbering, $\mu : \mathcal{P} \to \{1, 2, ..., |\mathcal{P}|\}$, is defined recursively by

(1) $\mu(\mathbf{0}) = 1$, and
(2) given $\mu^{-1}\{1, 2, ..., n-1\}$ for $n - 1 \geq 1$, let

$$r_0 = \max\{r : k_r = |\mu^{-1}\{1, 2, ..., n-1\} \cap \mathcal{P}_r| \text{ and } |\partial_r(\mu_r^{-1}\{1, 2, ..., k_r + 1\})| \leq |\mu^{-1}\{1, 2, ..., n-1\} \cap \mathcal{P}_{r-1}|\}$$

and then $\mu^{-1}(n) = \mu_{r_0}^{-1}(k+1)$.

Theorem 8.6 *If $\mathcal{P}$ is Macaulay then it has nested solutions for the MRI problem, i.e. the maximum weight ideal problem where the weight of an element is its rank. The rank-greedy Macaulay numbering, μ, is optimal.*

Proof If not, then $\exists S \in \mathfrak{I}(\mathcal{P})$, the set of ideals of $\mathcal{P}$, $|S| = k$, such that $r(S) = \sum_{x \in S} r(x) > r(\mu^{-1}\{1, 2, ..., k\})$. Let S be one such that maximizes $r(S)$ and minimizes $\mu(S)$ (over all those that maximize $r(S)$). Also, let a be the member of $\mu^{-1}\{1, 2, ..., k\} - S$ with minimum μ-value and b be the member of S with maximum μ-value. Then $\mu(a) \leq k < \mu(b)$. Also, $S \cap \mathcal{P}_r = \mu_r^{-1}\{1, 2, ..., k_r\}$ where $k_r = |S \cap \mathcal{P}_r|$, otherwise replacing $S \cap \mathcal{P}_r$ in each rank by $\mu_r^{-1}\{1, 2, ..., k_r\}$ would produce an ideal of the same cardinality and total rank but lower μ-value. Then $r(a) \geq r(b)$, since if $r(a) < r(b)$, look at $\overleftarrow{b} = \{x \in \mathcal{P} : x \leq b\}$, the ideal generated by b. Suppose that c is the member of $\overleftarrow{b} \cap \{x \in \mathcal{P} : \mu(x) > \mu(a)\}$ which minimizes $\mu(x)$. $r(a) < r(c)$ since $\overleftarrow{b} \subseteq S$ so $\overleftarrow{b} \cap \mathcal{P}_r \subseteq S \cap \mathcal{P}_r = \mu_r^{-1}\{1, 2, ..., k_r\}$ implying that $\forall x \in \overleftarrow{b} \cap \mathcal{P}_r$, $\mu(x) < \mu(a)$. Any members of $\overleftarrow{b}$ of rank lower than $r(a)$ must also have lower μ-value since μ is an extension of $\mathcal{P}$. But then $y \in \partial(c) \Rightarrow \mu(y) < \mu(a)$ which contradicts the definition of μ as rank-greedy. Since $r(a) \geq r(b)$ and $\mu(a) < \mu(b)$ we have

$$r(S - \{b\} + \{a\}) \geq r(S), \text{ and}$$
$$\mu(S - \{b\} + \{a\}) < \mu(S),$$

contradicting the definition of S. □

Our Theorem 8.6 above is essentially Theorem 8.3.1 of Engel's monograph [**34**]. The proof, also essentially the same as Engel's, works for all weights which are monotone increasing and constant on ranks. However, one must take care in comparing this exposition with Engel's because our definition of *compressed* is a bit different, corresponding more to Engel's term, "*j-compressed* for all *j*."

Corollary 8.2 *The Kruskal–Katona theorem (Theorem 8.2) implies that the EIP on Q_d has nested solutions and the optimal numbering is lexicographic (see Section 3.3.5 of Chapter 3).*

Proof The stability order of Q_d wrt complementation of the ith coordinate, $1 \leq i \leq d$, is $\mathcal{B}_d$. Thus the EIP, reduces to an MRI problem on $\mathcal{B}_d$. □

More generally, but with essentially the same proof, we have

Corollary 8.3 *The Clements–Lindström theorem (Theorem 8.3) implies Lindsey's Theorem (Theorem 6.4 of Chapter 6) that the EIP on the Hamming graph, $K_{n_1} \times K_{n_2} \times ... \times K_{n_d}$, has nested solutions and if $n_1 \geq n_2 \geq ... \geq n_d$, the optimal numbering is lexicographic.*

These corollaries affirm the intimate connection between the Macaulay property and the graphical isoperimetric problem, EIP, which we have addressed from Chapter 1.

8.2 Steiner operations for MSP

Now we extend the definition of Steiner operation in a way which brings fundamental insight to all problems we have studied. Before doing that, however, we must extend the definition of a combinatorial complex, $\mathcal{F}$, to include the possibility that $\mathcal{F}_r$, the set of faces of rank r, be partially ordered. The MSP extends to these new structures in the obvious way: calculate $\min_{\substack{S \in \mathcal{I}(\mathcal{F}_r) \\ |S|=k}} |\partial_r(S)|$. This additional structure must be respected by the boundary functions, i.e. $\forall S \in \Im(\mathcal{F}_r), \partial(S) \in \Im(\mathcal{F}_{r-1})$.

Then a *Steiner operation* $\phi : \mathcal{F} \to \mathcal{G}$, from a combinatorial complex, $\mathcal{F}$, to a combinatorial complex, $\mathcal{G}$ such that $<_{\mathcal{F}_r} \subseteq <_{\mathcal{G}_r}$, i.e. the partial order on $\mathcal{G}_r$, is an extension of that on $\mathcal{F}_r$ (so $\Im(\mathcal{G}_r) \subseteq \Im(\mathcal{F}_r)$), is comprised of maps $\phi_r : \Im(\mathcal{G}_r) \to \Im(\mathcal{F}_r)$, such that $\forall S, T \in \Im(\mathcal{G}_r)$

(1) $|\phi_r(S)| = |S|$
(2) $\partial(\phi_r(S)) \subseteq \phi_{r-1}(\partial(S))$
(3) If $S \subseteq T$, then $\phi_r(S) \subseteq \phi_r(T)$

(recall Theorem 2.3 of Chapter 2).

Theorem 8.7 *If $\phi : \mathcal{F} \to \mathcal{G}$ is a Steiner operation, then $\min_{\substack{S \in \Im(\mathcal{F}_r) \\ |S|=k}} |\partial(S)| = \min_{\substack{S \in \Im(\mathcal{G}_r) \\ |S|=k}} |\partial(S)|$ and the MSP on $\mathcal{F}$ reduces to that on $\mathcal{G}$.*

Proof

$$\begin{aligned} \left|\partial\left(\phi_r(S)\right)\right| &\leq \left|\phi_{r-1}(\partial(S))\right|, \text{ by (2)} \\ &= |\partial(S)|, \text{ by (1).} \end{aligned}$$

□

Thus Steiner operations preserve the MSP. It is the author's opinion that the inclusion between the intertwining operations of part (2) in the definition of Steiner operation for combinatorial complexes is fundamental in the study of combinatorial isoperimetric inequalities.

Theorem 8.8 *If $\mathcal{F}$,$\mathcal{G}$ are combinatorial complexes and $\phi : \mathcal{F} \to \mathcal{G}$ a Steiner operation, then $\phi^* : \mathcal{F}^* \to \mathcal{G}^*$, where $(\mathcal{F}^*)_r = (\mathcal{F}_r)^*$ and $\varphi_r^*(S) = \left(\varphi_r(S^c)\right)^c$.*

Theorem 8.9 *$\mathcal{F}$ is Macaulay iff there is a Steiner operation $\phi : \mathcal{F} \to \mathcal{T}$ where $\forall r$, $\mathcal{T}_r$ is totally ordered.*

Corollary 8.4 *$\mathcal{F}^*$ is Macaulay iff $\mathcal{F}$ is Macaulay.*

8.2.1 Stabilization

If we have the Hasse diagram of a ranked poset, $\mathcal{P}$, represented in $\mathbb{R}^d$, a *stabilizing reflection*, $\mathcal{R}$, of $\mathcal{P}$, is a reflection (of $\mathbb{R}^d$) which

(1) acts as an automorphism of $\mathcal{P}$, taking elements of $\mathcal{P}$ to elements of $\mathcal{P}$ and edges (of the Hasse diagram) to edges (preserving direction), and
(2) if $x \in \mathcal{P}$ and $y, z \in \partial(x)$, with y and z on different sides of the fixed hyperplane of $\mathcal{R}$, (i.e. $\|x - p\| < \|\mathcal{R}(x) - p\|$ and $\|y - p\| > \|\mathcal{R}(y) - p\|$) then $\mathcal{R}(x) = x$.

This definition extends that of Section 3.2.4 of Chapter 3. We have already extended the definition of stabilization (Section 5.6 of Chapter 5), but this is quite different. There may be a common extension of both but we shall not pursue it here. The Hasse diagram of the face lattice of a convex solid in $\mathbb{R}^d$ may be represented with a vertex $\left(\overline{f}, 1 + \dim(f)\right) \in \mathbb{R}^{d+1}$ for a face, f, where $\overline{f}$ is the centroid of f. One then sees that the reflective symmetries of regular solids become stabilizing symmetries of the Hasse diagram of its face lattice.

Theorem 8.10 *$Stab_{\mathcal{R},p}$ is a Steiner operation.*

Proof We must show that for all $S, T \subseteq \mathcal{P}_r$,

(1) $\left|Stab_{\mathcal{R},p}(S)\right| = |S|$,
(2) $\partial\left(Stab_{\mathcal{R},p}(S)\right) \subseteq Stab_{\mathcal{R},p}(\partial(S))$,

(3) If $S \subseteq T$ then $Stab_{\mathcal{R},p}(S) \subseteq Stab_{\mathcal{R},p}(T)$.

(1) and (3) are easy. To prove (2) we must show that

$$y \in \partial\left(Stab_{\mathcal{R},p}(S)\right) \text{ implies } y \in Stab_{\mathcal{R},p}(\partial(S)).$$

We note that

$$y \in \partial\left(Stab_{\mathcal{R},p}(S)\right) \Leftrightarrow \exists x \in Stab_{\mathcal{R},p}(S) \;\&\; y \in \partial(x)$$

and proceed by cases:

Proof

$x \in S$ and $\|x - p\| < \|\mathcal{R}(x) - p\|$: $\Rightarrow y \in \partial(S)$ and $\|y - p\| \leq \|\mathcal{R}(y) - p\|$
$\Rightarrow y \in Stab_{\mathcal{R},p}(\partial(S))$.
$x \in S$ and $\|x - p\| = \|\mathcal{R}(x) - p\|$, (i.e. $\mathcal{R}(x) = x$): $\Rightarrow y \in \partial(S)$ and then $\|y - p\| \leq \|\mathcal{R}(y) - p\|$
$\Rightarrow y \in Stab_{\mathcal{R},p}(\partial(S))$ or $\|y - p\| > \|\mathcal{R}(y) - p\|$
$\Rightarrow \mathcal{R}(y) \in \partial(S)$
$\Rightarrow y \in Stab_{\mathcal{R},p}(\partial(S))$.
$x \in S$ and $\|x - p\| > \|\mathcal{R}(x) - p\|$: $\Rightarrow y \in \partial(S)$ and $\|y - p\| \geq \|\mathcal{R}(y) - p\|$
$\Rightarrow \mathcal{R}(y) \in \partial(S)$
$\Rightarrow y \in Stab_{\mathcal{R},p}(\partial(S))$.
$x \notin S$ and $\|x - p\| < \|\mathcal{R}(x) - p\|$: $\Rightarrow \mathcal{R}(x) \in S$
$\Rightarrow \mathcal{R}(y) \in \partial(S)$ and $\|y - p\| \leq \|\mathcal{R}(y) - p\|$
$\Rightarrow y \in Stab_{\mathcal{R},p}(\partial(S))$.

□

As in Section 3.2 of Chapter 3, given multiple stabilizing symmetries, $\mathcal{R}_1, \mathcal{R}_2, \ldots, \mathcal{R}_k$ of $\mathcal{P}$, we need only consider stable sets in solving the MSP, and they are exactly the ideals in the stability order on $\mathcal{P}_r$.

Exercise 8.7 *Calculate the stability order for $\mathcal{B}_{d,r}, 0 \leq r \leq d$, wrt all $\mathcal{R}_{i,j}$, $1 \leq i < j \leq d$.*

Exercise 8.8 *Calculate the stability order for $\mathcal{Q}_{d,r}, 0 \leq r \leq d$, wrt all $\mathcal{R}_i$, $1 \leq i \leq d$ and $\mathcal{R}_{i,j}$, $1 \leq i < j \leq d$.*

Theorem 8.11 *The MSP on the face lattice, $\mathcal{F}$, of a regular solid reduces to the MWI on $\mathcal{S}(\mathcal{F}_i)$, the stability order of $\mathcal{F}_i$, $1 \leq i \leq d$.*

Proof The only thing remaining is to show that there are weights, $\#(x)$, for $x \in \mathcal{F}$, such that for all ideals, S, in $\mathcal{S}(\mathcal{F}_i)$,

$$|\partial(S)| = \sum_{x \in I} \#(x).$$

Since $\mathcal{F}$ is the face lattice of a regular convex polytope, whose group of symmetries, W, is Coxeter, $y \in \mathcal{F}_{i-1}$ contributes 1 to $|\partial(S + \{x\})| - |\partial(S)|$ iff x is minimal wrt the action of $W_y = \{g \in W : g(y) = y\}$ on $\mathcal{F}_i$. y^*, the face of $\mathcal{F}^*$, corresponding to y, is a regular solid with W_y as its (Coxeter) group of symmetries. W_y acts transitively on the members of $\mathcal{F}_i$ containing y, inducing a stability order which has a unique minimal element (see Theorems 5.4 and 5.5 of Chapter 5). Thus

$$\#(x) = \left| \left\{ y \in \partial(x) : \|\overline{x} - p\| = \min_{z \in \partial^*(y)} \|\overline{z} - p\| \right\} \right|.$$

□

Theorem 8.12 *If S is an ideal of $\mathcal{S}(\mathcal{F}_i)$, then $\partial(S)$ is an ideal of $\mathcal{S}(\mathcal{F}_{i-1})$.*

Proof $y \in \partial(S) \Leftrightarrow \exists y' \in S$ and $y = \partial(y')$ so we need to show that $x <_S y \in \partial(S) \Rightarrow \exists x' \in S$ and $x \in \partial(x')$: $x <_S y \Leftrightarrow \exists \mathcal{R}_{i_1}, \mathcal{R}_{i_2}, ..., \mathcal{R}_{i_n}$ such that $x_0 = x$, $x_1 = \mathcal{R}_{i_1}(x_0), x_2 = \mathcal{R}_{i_2}(x_1), ..., x_n = \mathcal{R}_{i_n}(x_{n-1}) = y$ with $\|x_0 - p\| < \|x_1 - p\| < ... < \|x_n - p\|$. Then let $x'_n = y'$, so $\partial(x'_n) = x_n$, and $x'_{n-1} = \mathcal{R}_{i_n}(x'_n)$. Then $x'_{n-1} \leq_S x'_n$ and $x_{n-1} = \partial(x'_{n-1})$ by the definition of stabilization. Continuing in this way, we define $x'_0 \leq_S x'_1 \leq_S ... \leq_S x'_n$ such that $x_i \in \partial(x'_i), 0 \leq i \leq n$. Therefore, $x'_0 \in S$ and $x = x_0 \in \partial(S)$. □

8.2.2 Compression

The following theorem generalizes Theorem 4.3 of Chapter 4.

Theorem 8.13 *If $\mathcal{U}$ and $\mathcal{V}$ are ranked posets and $\mathcal{U}$ is Macaulay, then compression on $\mathcal{U} \times \mathcal{V}$ is a Steiner operation for MSP.*

Proof For $S \subseteq (\mathcal{U} \times \mathcal{V})_{r_0}$, $Comp(S) = \bigcup_{y \in \mathcal{V}} \left(\mu^{-1}_{r_0 - r(y)} \{1, 2, ..., k_y\} \times \{y\} \right)$ where for $y \in \mathcal{V}$, $k_y = |S \cap (\mathcal{U} \times \{y\})|$. Then

(1) $|Comp(S)| = |S|$,
(2) $\partial(Comp(S)) \subseteq \partial(S)$,
(3) $S \subseteq T \Rightarrow Comp(S) \subseteq Comp(T)$.

(1) and (3) follow from the definition of compression but (2) requires the Macaulay properties of $\mathcal{U}$. □

If $\mathcal{P}$ is factorable as a product of subcomplexes, $\mathcal{U}_1 \times \mathcal{V}_1, \mathcal{U}_2 \times \mathcal{V}_2, ..., \mathcal{U}_d \times \mathcal{V}_d$ so that $\mathcal{U}_1, \mathcal{U}_2, ..., \mathcal{U}_d$ are Macaulay posets with rank-greedy numberings, $\mu^{(1)}, \mu^{(2)}, ..., \mu^{(d)}$, and $\mathcal{P}$ has a numbering μ, consistent with $\mu^{(1)}, \mu^{(2)}, ..., \mu^{(d)}$, then, as before, the compressions define compressibility orders on the $\mathcal{P}_i$, $0 \leq i \leq d$, and we need only consider compressed sets (ideals in the compressibility order) when solving the MSP on $\mathcal{P}_i$.

8.2.3 Lindström's theorem

Following Kruskal's suggestion that there should be an analog of the Kruskal–Katona theorem for $\mathcal{Q}_d$, the face lattice of the d-cube, B. Lindström published a solution, showing that $\mathcal{Q}_d$ is Macaulay [74] (although there was a flaw in the proof). The first challenge, when solving the MSP on some face lattice $\mathcal{F}$, is to find the Macaulay numberings of the $\mathcal{F}_r$. One way to ease the process is to utilize Theorem 8.6 and solve the maximum rank ideal problem (MRI) on $\mathcal{F}$ first. If that has nested solutions, and $\mathcal{F}$ is indeed Macaulay, then the optimal numbering, μ, should be a rank-greedy Macaulay numbering for $\mathcal{F}$. In that case $\mu_r = \mu|\mathcal{F}_r$, the restriction of μ to $\mathcal{F}_r$, will be a Macaulay numbering. For $\mathcal{Q}'_1 = \mathcal{Q}_1 - \{\mathbf{0}\}$ (the empty face is deleted), whose Hasse diagram is shown in Fig. 8.2, there is only one stable numbering: $\lambda(0) = 1$, $\lambda(1) = 2$ and $\lambda(\omega) = 3$, so it must maximize rank. $\mathcal{Q}'_2 = \mathcal{Q}_2 - \{\mathbf{0}\} = (\mathcal{Q}'_1)^2$ has the compressibility order shown in Fig. 8.3.

Note that the shadow function on the compressed sets of $\mathcal{Q}'_2$ is not additive since 0ω and $\omega 0$ are both minimal elements of rank 1 and their shadows both contain 00 (See Lemma 6.2 of Section 6.1 in Chapter 6). There are several stable (wrt the obvious symmetry) numberings of $\mathcal{Q}'_2$ and somewhat surprisingly, the lexicographic numbering is not optimal (remember we are maximizing rank, which is the number of ω's in x). An optimal numbering is given by the following table:

x	00	01	0ω	10	$\omega 0$	11	1ω	$\omega 1$	$\omega\omega$
$\lambda(x)$	1	2	3	4	5	6	7	8	9

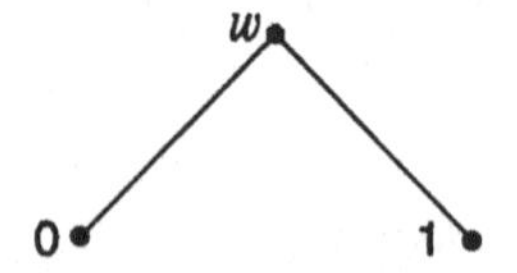

Fig. 8.2 Hasse diagram of $\mathcal{Q}'_1$.

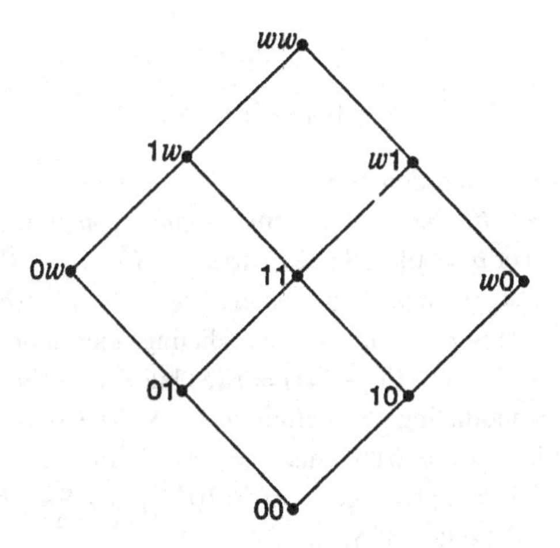

Fig. 8.3 Compressibility order of $\mathcal{Q}'_2$.

Assuming that the MWI problem on $\mathcal{Q}'_d$ (weight being rank) has nested solutions, we are led to the following order, first defined by Lindström: For $x, y \in \mathcal{Q}'_d = (\mathcal{Q}_1 - \{\mathbf{0}\})^d$, $\lambda(x) < \lambda(y)$ if

(1) $x|_{\omega=1} <_{lex} y|_{\omega=1}$ or
(2) $x|_{\omega=1} = y|_{\omega=1}$ and $x <_{lex} y$ (with $0 < 1 < \omega$).

Theorem 8.14 *For each k, $0 \le k \le 3^d$, $\lambda^{-1}\{1, 2, ..., k\}$ maximizes $r(S) = \sum_{x \in S} r(x)$ over all ideals, S, of $\mathcal{Q}'_d = \mathcal{Q}_d - \{\mathbf{0}\}$ with $|S| = k$.*

Proof By induction on d. We have already verified it for $d = 1, 2$. Assume it true for $d - 1 \ge 2$ and apply compression. Assume that S maximizes $r(S)$ over all ideals of $\mathcal{Q}_d - \{\mathbf{0}\} = (\mathcal{Q}_1 - \{\mathbf{0}\})^d$ with $|S| = k$ and, if there are several, minimizes $\lambda(S) = \sum_{x \in S} \lambda(x)$. If $S \ne \lambda^{-1}\{1, 2, ..., k\}$, let a be the member of $\lambda^{-1}\{1, 2, ..., k\} - S$ which minimizes $\lambda(x)$ and b be the member of S which maximizes $\lambda(x)$. Then $\lambda(a) < \lambda(b)$. Note that a and b, being incomparable wrt C, the compressibility order, are distinct in all components. There are then two cases:

(1) Neither a nor b has a 0 in any component. Then since $\lambda(a) < \lambda(b)$, $a_1 = 1$ and $b_1 = \omega$. If for any $i > 1$, $a_i = 1$ and $b_i = \omega$, it would imply that $a <_C b$, so $a = 1\omega^{d-1}$ and $b = \omega 1^{d-1}$. But then

$$r(a) = d - 1 > 1 = r(b)$$

and

$$r(S - \{b\} + \{a\}) > r(S),$$

contradicting the definition of S.

(2) Either a or b has a 0 in some component. Let $i_0 = \min\{i : a_i = 0 \text{ or } b_i = 0\}$. Since $\lambda(a) < \lambda(b)$, $a_{i_0} = 0$ and $b_{i_0} = 1$ or ω. Again, $\forall i \neq i_0, a_i > b_i$, or else $a <_C b$. If $r(a) > r(b)$ then $r(S - \{b\} + \{a\}) > r(S)$ again contradicting the definition of S. If $r(a) = r(b)$ then $r(S - \{b\} + \{a\}) = r(S)$ but $\lambda(S - \{b\} + \{a\}) < \lambda(S)$, once again contradicting the definition of S. The only possibility for $r(a) < r(b)$ is that $a = 01^{d-1}$ and $b = \omega 0^{d-1}$. But then a is succeeded by $a' = 01^{d-2}\omega$, b is preceded by $b' = 10^{d-1}$, and $S - \{b, b'\} + \{a, a'\}$ contradicts the definition of S.

□

One of the advantages of working with the MRI problem is that the rank function is, by definition, additive. The shadow function is not additive, even on (one-dimensionally) compressed sets, but we do have the following

Lemma 8.1 *If $S, S + \{x\} \subseteq \mathcal{Q}_{d,r}$ are two-dimensionally compressed wrt λ, then*

$$\begin{aligned}\#(x) &= |\partial(S + \{x\})| - |\partial(S)| \\ &= \left|\{x_i = \omega : \forall j > i, x_j = \omega\}\right| \\ &\quad + \left|\{x_i = \omega : \forall j > i, x_j \neq 1\}\right|.\end{aligned}$$

Thus $|\partial(S)| = \sum_{x \in S} \#(x)$, an additive function.

Exercise 8.9 *Prove Lemma 8.1.*

Theorem 8.15 *(Lindström* **[74]***) $\mathcal{Q}_d$ is a Macaulay poset and Lindström's numbering, λ, is optimal (rank-greedy, so $\lambda_r = \lambda|\mathcal{Q}_{d,r}$).*

Proof By induction on d:

$d = 1$: There is only one stable numbering for $r = 0, 1$ and that is λ_r.

$d = 2$: For $r = 0, 2$ there is only one stable numbering, λ_r. For $r = 1$ the stability order is shown in (Fig. 8.4). It has an extra stabilizing symmetry (for its MWI problem) and therefore only one stable numbering.

$d \geq 3$: Let $S_k \subseteq \mathcal{Q}_{d,r}$ be a compressed set which minimizes $|\partial(S)|$ over all sets of cardinality k and also minimizes $\lambda_r(S) = \sum_{x \in S} \lambda_r(x)$. If $S_k \neq$

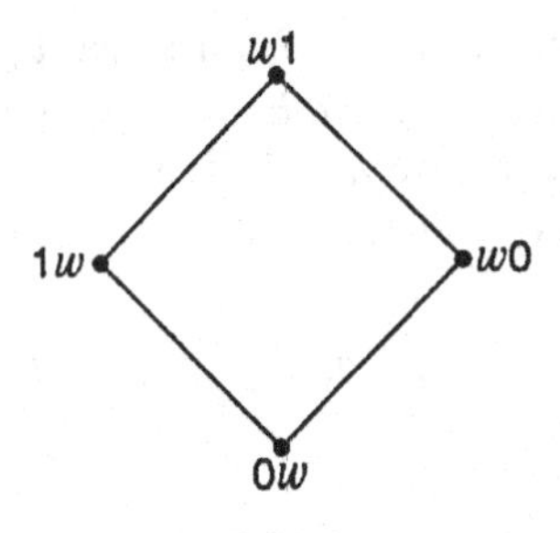

Fig. 8.4.

$\lambda_r^{-1}\{1, 2, ..., k\}$, then $\exists a, b \in \mathcal{Q}_{d,r}$ such that

$$\lambda_r(a) = \min\{\lambda_r(x) : x \notin S_k\} \\ < \max\{\lambda_r(x) : x \in S_k\} = \lambda_r(b).$$

Our strategy, as in previous such arguments, is to show that $S_k - \{b\} + \{a\}$ contradicts the definition of S_k. All we need is that $\#(a) \leq \#(b)$, which brings us to the combinatorial heart of the argument: Note that a and b are incomparable wrt $\mathcal{C}$ so $\forall i, a_i \neq b_i$. By the pigeonhole principle, then, $r \leq d/2$. $r = 0$ is trivial since $\forall a \in \mathcal{Q}_{d,0}\ \#(a) = 0$. Then we consider two cases:

(1) $\forall i, a_i \neq 0$ *and* $b_i \neq 0$: Then $a_1 = 1$ and $b_1 = \omega$. d must be even and $r = d/2$. If $\exists 1 < i < j$ with $a_i = 1$ and $a_j = \omega$, then

$$a <_{\mathcal{C}} \mathcal{R}_{ij}(a) <_{\mathcal{C}} b.$$

The only other possibility is that $a = 1\omega^{d/2}1^{d/2-1}$, $b = \omega 1^{d/2}\omega^{d/2-1}$but then $\#(a) = 0 < \#(b)$.

(2) $\exists i, a_i = 0$ or $b_i = 0$: Let $i_0 = \min\{i : a_i = 0 \text{ or } b_i = 0\}$. Then $a_{i_0} = 0$. If $i, j \neq i_0$ and $a_i = \omega = b_j$, then $i < j$, since if $i > j$ then $a_j \neq \omega$ so

$$a <_{\mathcal{C}} \mathcal{R}_{ij}(a) <_{\mathcal{C}} b.$$

Also, $\forall j \neq i_0, a_j \neq 0$, since if $a_j = 0$, then $b_j \neq 0$. If $b_j = 1$ then

$$a <_{\mathcal{C}} a + \delta^{(j)} <_{\mathcal{C}} b.$$

Or if $b_j = \omega$ then, as before, if $a_i = \omega$ (and there must be such i since $r > 0$) then $i < j$ and

$$a <_{\mathcal{C}} \mathcal{R}_{ij}(a) + \delta^{(i)} <_{\mathcal{C}} b.$$

Therefore if there exists $j \neq i_0$ such that $b_j = \omega$, then $\#(a) = 0 \leq \#(b)$. This means that $r = 1$ and $b_{i_0} = \omega$. Then if $i_0 > 1$, $a_1 = \omega$ which implies that $\#(a) = 0$ since $d \geq 3$. The only remaining possibility for $\#(a) > \#(b)$ is $a = 01^{d-2}\omega$ and $b = \omega 0^{d-2}1$ but then

$$a <_{\mathcal{C}} \mathcal{R}_{d-1,d}(a) <_{\mathcal{C}} b.$$

□

8.2.4 Extensions of Lindström's theorem

8.2.4.1 The theorems of Leeb and Bezrukov–Elsässer

K. Leeb [**71**] showed that Lindström's theorem could be extended to products of "star" posets (also called "hands" in the literature). Finally, Bezrukov and Elsässer [**19**] extended Leeb's result to products of "spiders" and showed there could be no further extensions in this direction.

For $m, n \in \mathbb{Z}^+$, an $m \times n$ *spider poset,* $\mathcal{S}p(m, n)$ has elements

$$Sp = \{\omega\} + \{1, 2, ..., m\} \times \{1, 2, ..., n\},$$

ω being a maximum element, and otherwise

$$(i, j) < (i, j+1) \ \forall j \in \{1, 2, ..., n-1\}, \ 1 \leq i \leq m.$$

See Fig. 8.5.

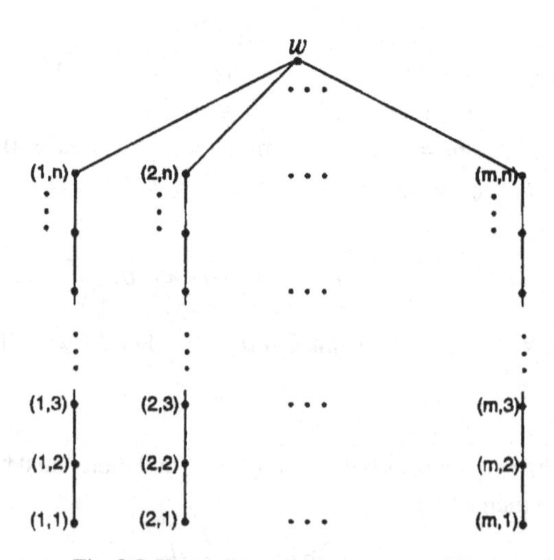

Fig. 8.5 Hasse diagram of an $m \times n$ spider.

Note that $\mathcal{B}_d \simeq (\mathcal{S}p\,(1,1))^d$ and $\mathcal{Q}_d \simeq (\mathcal{S}p\,(2,1))^d + \{\mathbf{0}\}$. The case $n = 1$ is what Engel [**34**] calls a *star*.

Theorem 8.16 [**19**] *All powers of a spider poset, $(\mathcal{S}p(m,n))^d$, are Macaulay.*

The theorem probably extends to products of spiders with variable number of legs, $\mathcal{S}p(m_1, n) \times \mathcal{S}p(m_2, n) \times \ldots \times \mathcal{S}p(m_d, n)$, but it is necessary that the lengths of the legs be the same. The proof of Bezrukov–Elsässer [**19**] is complex, and so far we have not been able to simplify it.

8.2.4.2 The theorems of Vasta and Leck

Another extension of Lindström's theorem was announced recently. Again, the proofs are complicated and probably close to minimal anyway, so we shall only state the theorem. First a definition: If $\mathcal{F}$ and $\mathcal{G}$ are combinatorial complexes (ranked posets with **0** and **1**), their *rectangular product*, $\mathcal{F}\square\mathcal{G}$, is defined by

$$\mathcal{F}\square\mathcal{G} = (\mathcal{F} - \{\mathbf{0}\}) \times (\mathcal{G} - \{\mathbf{0}\}) + \{\mathbf{0}\}.$$

Rectangular product is natural for the face lattices of convex polytopes since the face lattice of a product of convex polytopes is the rectangular product of their face lattices. It is also the natural extension of the product of graphs to higher dimensional complexes. In the early 1980s the author suggested that the rectangular products of simplices could be Macaulay. This extends Lindström's theorem because $\mathcal{Q}_d$ is the rectangular product of intervals ($\mathcal{Q}_1$'s, which are also 1-simplices). It also extends Lindsey's theorem (Theorem 6.4 of Chapter 6). In the early 1990s A. Sali studied various optimization problems, including MSP, on the submatrices of an $m \times n$ matrix, partially ordered by containment [**86**]. This poset is isomorphic to $\mathcal{B}_m\square\mathcal{B}_n$, the pairwise rectangular product of simplices. In 1998 J. Vasta [**89**] solved the maximum rank ideal (MRI) problem for rectangular products of simplices, showing that they had nested solutions, and then U. Leck [**70**] completed the process of proving that they are Macaulay. Note that the rectangular product does not commute with duality (see Exercise 8.2).

Theorem 8.17 [**70**] *The rectangular product of simplices, $\mathcal{B}_{n_1}\square\mathcal{B}_{n_2}\square\ldots\square\mathcal{B}_{n_d}$, is Macaulay.*

The (rank-greedy) Macaulay ordering, μ, of

$$\begin{aligned}&\mathcal{B}_{n_1}\square\mathcal{B}_{n_2}\square\ldots\square\mathcal{B}_{n_d} - \{\mathbf{0}\}\\ &= \left(\mathcal{B}_{n_1} - \{\mathbf{0}\}\right) \times \left(\mathcal{B}_{n_2} - \{\mathbf{0}\}\right) \times \ldots \times \left(\mathcal{B}_{n_d} - \{\mathbf{0}\}\right)\end{aligned}$$

is defined recursively as follows. If $d = 1$ then μ is just lexicographic order. If $d > 1$ and $x, y \in \left(\mathcal{B}_{n_1} - \{\mathbf{0}\}\right) \times \left(\left(\mathcal{B}_{n_2} - \{\mathbf{0}\}\right) \times \ldots \times \left(\mathcal{B}_{n_d} - \{\mathbf{0}\}\right)\right)$ such that

$x = (a, b)$, $y = (c, d)$ with $a, c \in \{0, 1\}^{n_1} - \{0^{n_1}\}$, $b, d \in \left(\mathcal{B}_{n_2} - \{\mathbf{0}\}\right) \times \ldots \times \left(\mathcal{B}_{n_d} - \{\mathbf{0}\}\right)$ and

$$a_i = \begin{cases} 0 & \text{if } i < k \\ 1 & \text{if } i = k, \end{cases}$$

$$c_i = \begin{cases} 0 & \text{if } i < \ell \\ 1 & \text{if } i = \ell, \end{cases}$$

then $x <_\mu y$ if

(1) $k > \ell$, or
(2) $k - \ell$ and $b <_\mu d$, or
(3) $k = \ell$ and $b = d$ and $a <_{lex} c$.

Note that if we add a null face to the product of spider posets, it becomes a rectangular product of symmetric and self-dual complexes,

$$\begin{aligned} &\mathcal{S}(m_1, n) \times \mathcal{S}(m_2, n) \times \ldots \times \mathcal{S}(m_d, n) + \{\mathbf{0}\} \\ &= (\mathcal{S}(m_1, n) + \{\mathbf{0}\}) \square (\mathcal{S}(m_2, n) + \{\mathbf{0}\}) \square \ldots \square (\mathcal{S}(m_d, n) + \{\mathbf{0}\}). \end{aligned}$$

8.3 Scheduling problems

8.3.1 Profile scheduling

Scheduling is an important topic in operations research, the applied part of combinatorial optimization. *MathSci*, the online version of *Mathematical Reviews*, currently (August, 2001) lists 2620 articles under the keyword "scheduling" published since 1978, somewhere around 150 of them in the year 2000. There are many variants of the scheduling problem, and they are generally NP-hard, so we shall restrict our attention to a basic few. Suppose that the steps of a manufacturing process must be carried out in some serial order (on a single processor) subject to given precedence constraints. If each step takes a unit of time, and after its completion we must store its product for use in the steps which follow it directly, then it is the total storage time, the sum of the number of parts in storage during each unit of time of the process, which we wish to minimize. The precedence constraints will be represented by a partial order, $\mathcal{P} = (P, <)$, so that if $x < y$ then step x must be completed before step y. A *schedule* is a numbering, $\eta : P \to [n]$, $n = |P|$, which is an extension of $\mathcal{P}$ (if $x <_{\mathcal{P}} y$ then $\eta(x) < \eta(y)$). When the last successor, y, of x is completed, we need no longer store the product of x, and the time it spent in storage is

$\eta(y) - \eta(x)$. The total storage time will then be

$$\sum_{x \in P} \max_{x \lessdot y} (\eta(y) - \eta(x))$$

where "$x \lessdot y$" means that $x < y$ and there is no z such that $x < z < y$. In this case we say that y *covers* x. This functional is called the *profile of* η [**67**] and finding its minimum over all numberings of $\mathcal{P}$ is the *profile scheduling problem*. It is easy to see that $\forall k$, $\eta^{-1}[k]$ is an ideal of $\mathcal{P}$ and if for any ideal, S, $\Phi'(S) = \{x \in S : \exists\, y \notin S \,\&\, x \lessdot y\}$ then

$$\sum_{x \in P} \max_{x \lessdot y} (\eta(y) - \eta(x)) = \sum_{k=0}^{n} \left|\Phi'\left(\eta^{-1}\{1, 2, ..., k\}\right)\right|.$$

So the profile scheduling problem reduces to a minimum path problem on the ($|\Phi'|$-weighted) derived network of $\mathcal{P}$.

Corollary 8.5 *(of Theorem 4.2, Chapter 4) Hales numbering solves the profile scheduling problem for $\mathcal{B}_d$.*

Proof One-dimensional stabilization on $\mathcal{Q}_d$ reduces its vertex isoperimetric problem (VIP) to the scheduling VIP on $\mathcal{B}_d$. □

Corollary 8.6 *The generalized Hales numbering solves the profile scheduling problem for a product of total orders, $\mathcal{T}_{n_1} \times \mathcal{T}_{n_2} \times ... \times \mathcal{T}_{n_d}$.*

Proof This follows from the Chvatalova–Moghadam theorem, mentioned in the Comments of Chapter 6. □

The fact that the one-dimensional compressibility order of a product of paths, $P_{n_1} \square P_{n_2} \square ... \square P_{n_d}$ is $\mathcal{T}_{n_1} \times \mathcal{T}_{n_2} \times ... \times \mathcal{T}_{n_d}$ makes the latter very special. It is natural to ask if there are ranked posets which are not the compressibility orders of graphs that have generative nested solutions for VIP. For any such poset the optimal numbering would have to be rank-by-rank (generalized Hales) and dual-Macaulay, μ^*, within each rank. Of the known Macaulay posets, $\mathcal{Q}_d$, the complex of the d-cube is the most obvious candidate for this honor.

It is desirable to define the Hales–Lindström numbering in such a way that it is stable in order to apply Lemma 8.1. λ_r^*, the restriction of λ^* to $\mathcal{Q}_{d,r}$, is obviously not stable (i.e. an extension of the stability order on $\mathcal{Q}_{d,r}$), but rather surprisingly, λ_r itself is also not stable in general.

Exercise 8.10 *Find some $\mathcal{Q}_{d,r}$ for which λ_r is not stable.*

It may, of course, be stabilized and retain its optimality and we now do this. For $x, y \in \mathcal{Q}'_d = (\mathcal{Q}_1 - \{\mathbf{0}\})^d$, $\lambda_s(x) < \lambda_s(y)$ (s for stable) if

(1) $x|_{\omega=1} <_{lex} y|_{\omega=1}$ or
(2) $x|_{\omega=1} = y|_{\omega=1}$ and $x <_{lex} y$ (but this time with $0 < \omega < 1$).

Lemma 8.2 *λ_s is stable.*

Proof The stability order, $\mathcal{S}$, of $\mathcal{Q}_{d,r}$ is generated by pairs $x <_{\mathcal{S}} y$ for which

(1) $y = \mathcal{R}_i(x)$, $\mathcal{R}_i$ being the reflection which transposes 0 and 1 in the ith coordinate, with $x_i = 0$,
(2) $y = \mathcal{R}_{i,j}(x)$, $i < j$, $\mathcal{R}_{i,j}$ being the reflection which transposes the ith and jth coordinates, with
    (a) $x_i = 0$ and $x_j = 1$, or
    (b) $x_i = 0$ and $x_j = \omega$, or
    (c) $x_i = \omega$ and $x_j = 1$.

Consider then the relationship of $\lambda_s(x)$ and $\lambda_s(y)$ for any such generating pair:

(1) $y_i = 1$, so $x|_{\omega=1} <_{lex} y|_{\omega=1}$ and $\lambda_s(x) < \lambda_s(y)$,
    (a) again, $y_i = 1$, so $x|_{\omega=1} <_{lex} y|_{\omega=1}$ and $\lambda_s(x) < \lambda_s(y)$,
    (b) $y_i = \omega$ but $y_j = 0$, so $x|_{\omega=1} <_{lex} y|_{\omega=1}$ and $\lambda_s(x) < \lambda_s(y)$,
    (c) $y_i = 1$ and $y_j = \omega$, so $x|_{\omega=1} = y|_{\omega=1}$, $x <_{lex} y$ (remember $\omega < 1$ now) and $\lambda_s(x) < \lambda_s(y)$.

□

Exercise 8.11 *Show that λ_s works as well as λ in the proof of Lindström's theorem (8.16).*

Now we define the Hales-Lindström numbering on $\mathcal{Q}_d$: $x <_{HL} y$ iff

(1) $r(x) < r(y)$, or
(2) $r(x) = r(y)$ and $\overline{x} >_{\lambda_s} \overline{y}$, where

$$\overline{x}_i = \begin{cases} 1 & \text{if } x_i = 0 \\ 0 & \text{if } x_i = 1 \\ \omega & \text{if } x_i = \omega. \end{cases}$$

Lemma 8.3 *The Hales–Lindström numbering is stable.*

Exercise 8.12 *Use Bezrukov's poset tools (http://mcs.uwsuper.edu/sb/posets/) to verify that $\mathcal{Q}_2$ and $\mathcal{Q}_3$ do have generative nested solutions for the minimum (upper) shadow ideal problem.*

However, $\mathcal{Q}_4$ has too many ideals to solve the problem by brute force.

Exercise 8.13 *Calculate the stability order of $\mathcal{Q}_4$.*

Exercise 8.14 *Find a formula for the weight, $\#(x)$, of each $x \in \mathcal{S}(\mathcal{Q}_4)$, such that*

$$|\Phi(S)| = \sum_{x \in S} \#(x).$$

Exercise 8.15 *Use the results of the previous two exercises (and Bezrukov's poset tools) to show that $\mathcal{Q}_4$ does not have nested solutions for the minimum (upper) shadow ideal problem.*

8.3.2 Bandwidth scheduling and wirelength scheduling

If the definition of the profile of a numbering, η, above is modified to

$$\max_{\substack{x \in P \\ x \lessdot y}} (\eta(y) - \eta(x)),$$

we may call it the *bandwidth of* η, Minimizing it over all extensions of $\mathcal{P}$, would then be the *bandwidth scheduling problem.* Obviously,

$$\max_{\substack{x \in P \\ x \lessdot y}} (\eta(y) - \eta(x)) \geq \max_{0 \leq k \leq n} \left|\Phi\left(\eta^{-1}\{1, 2, ..., k\}\right)\right|,$$

$$\geq \max_{0 \leq k \leq n} \min_{\substack{S \in \mathcal{I}(\mathcal{P}) \\ |S=k|}} |\Phi(S)|$$

so that any poset, such as $\mathcal{B}_d$ or $\mathcal{T}_{n_1} \times \mathcal{T}_{n_2} \times ... \times \mathcal{T}_{n_d}$, whose VIP has generative nested solutions will have a nice solution for its bandwidth problem.

Similarly, minimizing

$$\sum_{\substack{x, y \in P \\ x \lessdot y}} (\eta(y) - \eta(x))$$

over all numberings, η, may be called the *wirelength scheduling problem.* As before,

$$\sum_{\substack{x, y \in P \\ x \lessdot y}} (\eta(y) - \eta(x)) = \sum_{k=0}^{n} \left|\Theta\left(\eta^{-1}\{1, 2, ..., k\}\right)\right|$$

$$\geq \sum_{k=0}^{n} \min_{\substack{S \in \mathcal{I}(\mathcal{P}) \\ |S=k|}} |\Theta(S)|.$$

Theorem 8.18 *Lexicographic numbering solves the wirelength scheduling problem on $\mathcal{B}_d$.*

Theorem 8.19 *Lindström's numbering solves the wirelength scheduling problem on $\mathcal{Q}_d$.*

Lemma 8.4 *For $S \in \mathcal{I}(\mathcal{Q}_d)$ with $|S| = k$,*

$$|\Theta(S)| = kd - 3\sum_{x \in S} r(x).$$

Proof By induction on k. It is trivially true for $k = 0$. Assume it is true for $k - 1 \geq 0$. Then if $|S| = k > 0$, we may choose $x_0 \in S$ and maximal wrt the partial order. Then $S - \{x_0\} \subset \mathcal{I}(\mathcal{Q}_d)$ and $|S - \{x_0\}| = k - 1$, so

$$\begin{aligned}|\Theta(S)| &= |\Theta(S - \{x_0\})| + (|\Theta(S)| - |\Theta(S - \{x_0\})|) \\ &= (k-1)d - 3\sum_{x \in S - \{x_0\}} r(x) + (d - 3r(x_0))\end{aligned}$$

by the induction hypothesis and the observation that the marginal contribution of x_0 to $|\Theta(S)|$ is $d - 3r(x_0)$. Each of the d components of x_0 which are not ω will contribute 1 to it and those that are ω will not only contribute 0, but will cancel two contributions from predecessors of x_0 (those gotten by replacing ω by 0 or 1). □

Proof (of Theorem 8.19). Thus minimizing wirelength can be accomplished by maximizing

$$r\left(\eta^{-1}\{1, 2, ..., k\}\right)$$

for each $k = 1, 2, ..., n = 3^d$, which λ has been shown to do. □

Exercise 8.16 *Show that Vasta's (rank-greedy Macaulay) numbering solves the wirelength scheduling problem on $\mathcal{B}_{n_1} \square \mathcal{B}_{n_2} \square ... \square \mathcal{B}_{n_d}$.*

Exercise 8.17 *Does Theorem 8.19 extend to products of spider posets, $\mathcal{S}(m_1, n) \times \mathcal{S}(m_2, n) \times ... \times \mathcal{S}(m_d, n)$?*

8.4 Comments

8.4.1 More MSPs

The minimum shadow problem on many interesting complexes remains unsolved:

(1) The (two-dimensional faces of the) exceptional regular solids in four dimentsions.
(2) Rectangular products of crosspolytopes (are they Macaulay?).
(3) Rectangular products of pentagonal tessellations of the projective plane (whose graphs are the products of Petersen graphs).
(4) Rectangular products of n-gons ($\mathbb{Z}_n$'s), $n \leq 5$.

One of the first applications of Bezrukov's poset tool (http://mcs.uwsuper.edu/sb/posets/) was to solve the EIP on the pairwise product of dodecahedral graphs and show that it does not have nested solutions.

8.4.2 Other complexes

There are other complexes, besides those generated by Coxeter groups, whose MSP has been studied, but results are sparse. $PG(d, q)$, the d-dimensional projective geometry over the Galois field of q elements, q a power of a prime, is one such. $PG(d, q)$ has a transitive symmetry group generated by "reflections", but these reflections do not give Steiner operations. That $PG(d, q)$ is a *normal* poset (see [**34**]) was one of the first fruits of the global approach to Sperner problems (see [**34**]). For the Sperner problem, being normal (also known as the LYM property) is analogous to being Macaulay for the MSP. The author had asked people about this problem for years before a visiting colleague, F. Hergert, suggested that Segre's theorem on ovals in $PG(2, 2^r)$ might give solutions of the MSP. In the resulting paper [**53**], we showed that unions of conics with a common nucleus are solutions of MSP for cardinalities $k = 1 + \left(2^l - 1\right)(2^r + 1)$, $0 \leq l \leq r$. Since these are nested and could be extended to nested solutions for all cardinalities when $r = 1, 2$, we conjectured that $PG(2, 2^r)$ should have nested solutions for the MSP for all r. P. K. Ure [**88**] validated our conjecture for $k = 2 + \left(2^l - 1\right)(2^r + 1)$, $0 \leq l \leq r$, but then showed by computer that $PG\left(2, 2^3\right)$ (which has $8^2 + 8 + 1 = 73$ points and 73 lines) does not have nested solutions.

Bezrukov and Blokhuis [**17**] found similar results for the linear lattice, $PG(d, 2)$. For each k, initial k-segments of lex order on c-dimensional subspaces have minimum 0-shadow (the total number of points in the subspace). The dual result also holds, of course, but they also showed by example that $PG(d, 2)$ is not Macaulay.

Another family of complexes ripe for investigation are Greene's posets of shuffles [**42**].

8.4.3 Bezrukov's equivalence principle

In [**16**] Bezrukov introduces a different connection between the EIP on a graph and the MRI on a ranked poset. He gives an algorithm for passing from a graph to a ranked poset whose MRI is essentially equivalent to the EIP on the original graph. The poset is called the *representing poset* of the graph. He proves that the representing poset of a product of graphs is the product of their representing posets so that the solution of the EIP on certain graph products can be deduced from a solution of the MRI on the corresponding product posets, which in turn can be obtained from known results on shadow minimization. One application of these ideas is the solution of the induced EIP on any product of trees.

8.4.4 Combinatorics since the 1960s

The growth of combinatorics since the 1960s has been amazing. When the *Journal of Combinatorial Theory* was established in 1966 (by F. Harary and G.-C. Rota) it was the first journal specializing in combinatorics. [**46**] appeared in Volume 1 of JCT and was the first to apply the term "isoperimetric" to optimization problems on graphs. In 1969 when Krukal's article [**66**] and the Clements–Lindström article [**29**] appeared, *JCT* was still the only journal devoted to combinatorics. In 1972, due to pressure from an increasing backlog of worthy papers, *JCT* split into *JCT-A* and *JCT-B*. The Combinatorics Net webpage now lists 59 journals specializing in combinatorics.

9

Morphisms for MWI problems

In Chapter 6, Section 6.2 we observed that for graphs having nested solutions for EIP, compression reduces the EIP on their product to a maximum weight ideal (MWI) problem on its compressibility order. Also, stabilization on the graph of any regular solid reduces both its EIP and VIP to MWI problems (on its stability order, but with different weights). The MWI problem, like the EIP and VIP, is NP-complete (see Section 6.2 of Chapter 6), but now we show that it has its own notion of morphism. MWI-morphisms extend, but are qualitatively different from, Steiner operations for EIP and VIP, in that the underlying functions may be many-one. They represent a divide-and-conquer strategy for solving MWI problems, but unlike the elementary applications of divide-and-conquer to sorting, etc. one must divide cleverly in order to conquer.

9.1 MWI-morphisms

9.1.1 Quotients

If **P** and **Q** are weighted posets and $\varphi : P \rightarrow Q$ a (many-one) onto function from which we hope to get a morphism for the MWI problem, what properties must it have? As in our previous considerations of this basic question, it must "preserve" the structures which define the MWI problem, i.e.

(1) Partial order: $x \leq y$ implies $\varphi(x) \leq \varphi(y)$;
(2) Weights: $\forall x \in Q$, $|x| = \left|\varphi^{-1}(x)\right|$ defines a "*cardinality function*", $|\cdot|$, on Q. Also, $\forall x \in Q$ there must be a list of weights, $\Delta(x, i)$, $1 \leq i \leq |x|$, such that $\forall S \in \mathfrak{I}\left(\varphi^{-1}(x)\right)$, the partial order on $\varphi^{-1}(x)$ being the one inherited from **P**, $\Delta(S) \leq \sum_{i \leq |S|} \Delta(x, i)$ and that inequality is sharp. This means that $\Delta(x, i) = MWI\left(\varphi^{-1}(x), \Delta; i\right) - MWI\left(\varphi^{-1}(x), \Delta; i-1\right)$ or

equivalently, $\Delta(x, \{1, ..., i\}) = MWI\left(\varphi^{-1}(x), \Delta; i\right)$ (see Definition 6.2.1 of Chapter 6).

Note that the set $\overline{Q} = \{(x, i) : x \in Q \;\&\; 1 \leq i \leq |x|\}$, implicit in the preceding paragraph, has cardinality $\left|\overline{Q}\right| = |P|$. It turns out that $\overline{Q}$ may be partially ordered in several different, but natural, ways. The most obvious, and the strongest, is $\overline{\mathbf{Q}}_{st}$, defined by $(x, i) \leq_{\overline{\mathbf{Q}}_{st}} (y, j)$ if $x <_{\mathbf{Q}} y$ or if $x = y$ and $i \leq j$. This partial order leads us to define a function

$$MinShadow(x, y; j) = \min\left\{\left|\overleftarrow{T} \cap \varphi^{-1}(x)\right| : T \in \mathfrak{I}\left(\varphi^{-1}(y)\right) \right.$$
$$\left. \& \; |T| = j\right\}.$$

Given MinShadow, we can define the weakest partial order, $\overline{\mathbf{Q}}_{wk}$, on $\overline{Q}$, by

$$(x, i) \leq_{\overline{\mathbf{Q}}_{wk}} (y, j)$$

if $x \leq_{\mathbf{Q}} y$ and $i \leq MinShadow(x, y; j)$. Clearly,

$$\leq_{\overline{\mathbf{Q}}_{wk}} \subseteq \leq_{\overline{\mathbf{Q}}_{st}},$$

so

$$\mathfrak{I}\left(\overline{\mathbf{Q}}_{st}\right) \mathfrak{I}\left(\overline{\mathbf{Q}}_{wk}\right)$$

which implies that

$$MWI\left(\overline{\mathbf{Q}}_{st}, \Delta; k\right) \leq MWI\left(\overline{\mathbf{Q}}_{wk}, \Delta; k\right)$$

$\forall k \in \mathbb{Z}_{+}$.

Now, given $x \in Q$ and $0 \leq i \leq |x|$, let $S_{x,i} \in \mathfrak{I}\left(\varphi^{-1}(x)\right)$ be a (any) solution of the MWI problem on $\varphi^{-1}(x)$ with $\left|S_{x,i}\right| = i$.

Lemma 9.1 *If $\varphi : P \rightarrow Q$ is order and weight preserving (as defined above), then*

(1) *For $S \in \mathfrak{I}\left(\overline{\mathbf{Q}}_{st}\right)$ and $x \in Q$, let $c_S, (x) = \max\{i : (x, i) \in S\}$. Then, by the definition of $\leq_{\overline{\mathbf{Q}}_{st}}$,*

$$\overline{\varphi}^{-1}(S) = \bigcup_{x \in Q} S_{x, c_S(x)} \in \mathfrak{I}(\mathbf{P}),$$

so $\left|\overline{\varphi}^{-1}(S)\right| = |S|$ and $\Delta\left(\overline{\varphi}^{-1}(S)\right) = \Delta(S)$. Therefore

$$MWI(\mathbf{P}, \Delta; k) \geq MWI\left(\overline{\mathbf{Q}}_{st}, \Delta; k\right)$$

$\forall k \in \mathbb{Z}_{+}$.

(2) *For* $S \in \mathfrak{I}(\mathbf{P})$ *and* $x \in Q$, *let* $c'_S(x) = \left|S \cap \varphi^{-1}(x)\right|$. *Then, by the definition of* $\leq_{\overline{\mathbf{Q}}_{wk}}$,

$$\overline{\varphi}(S) = \bigcup_{x \in Q} \left\{(x, i) : 1 \leq i \leq c'_S(x)\right\} \in \mathfrak{I}\left(\overline{\mathbf{Q}}_{wk}\right),$$

so $|\overline{\varphi}(S)| = |S|$ *and* $\Delta(\overline{\varphi}(S)) \geq \Delta(S)$. *Therefore*

$$MWI(\mathbf{P}, \Delta; k) \leq MWI\left(\overline{\mathbf{Q}}_{wk}, \Delta; k\right)$$

$\forall k \in \mathbb{Z}_+$.

The question then is, what additional conditions will make the inequalities of Lemma 9.1 sharp?

9.1.2 The main definitions

9.1.2.1 Strong MWI-morphisms

In order to ensure that the inequality of Lemma 9.1, part (1), is an equality, we would need to show that $\forall S \in \mathfrak{I}(\mathbf{P}),\ \exists S' \in \mathfrak{I}\left(\overline{\mathbf{Q}}_{st}\right)$ such that $\left|S'\right| = |S|$ and $\Delta\left(S'\right) \geq \Delta(S)$. If this is the case, we call $\varphi : \mathbf{P} \to \overline{\mathbf{Q}}_{st}$ a *strong MWI-morphism.* However, an effective demonstration of existence should be the result of an efficient algorithm. Such is the following (conditions (1) and (2) are just repeating the necessary conditions of Section 9.1.1)

Definition 9.1 *Let* $\mathbf{P}$ *be a poset with weight* Δ, $\mathbf{Q}$ *a poset and* $\varphi : P \to Q$ *a function. Then* φ *is a skeletal MWI-morphism,* $\varphi : \mathbf{P} \to \overline{\mathbf{Q}}_{st}$, *if*

(1) $\forall x, y \in \mathbf{P}$, $x \leq y$ *implies* $\varphi(x) \leq \varphi(y)$);
(2) φ *is weight-preserving:* $\forall x \in \mathbf{Q}$ *and* $0 \leq i \leq |x|$,

$$\Delta_{\overline{\mathbf{Q}}}(x, \{1, ..., i\}) = MWI\left(\varphi^{-1}(x), \Delta; i\right)$$

where $\varphi^{-1}(x)$ *inherits its partial order, cardinality and weight from* $\mathbf{P}$.

(3) $\forall x <_{\mathbf{Q}} y; \forall j, 0 < j \leq |y|; \forall i \geq MinShadow(x, y; j)$ *either*

(a) $i + j \leq |x|$ *and (dropping the subscript,* $\overline{\mathbf{Q}}$, *on* Δ)

$$\Delta(x, \{1, ..., i\}) + \Delta(y, \{1, ..., j\}) \leq \Delta(x, \{1, ..., i + j\})$$

or

(b) $i + j > |x|$ *and*

$$\Delta(x, \{1, ..., i\}) + \Delta(y, \{1, ..., j\}) \leq \Delta(x) + \Delta(y, \{1, ..., i + j - |x|\}).$$

Theorem 9.1 *A skeletal MWI-morphism is a strong MWI-morphism.*

Proof Given $S \in \mathfrak{I}(\mathbf{P})$ then, by Lemma 9.1, part (2), $S' = \overline{\varphi}(S) \in \mathfrak{I}(\overline{\mathbf{Q}}_{wk})$. If $S' \in \mathfrak{I}(\overline{\mathbf{Q}}_{st})$, we are done. If not, then $\exists (x, c_{S'}(x)), (y, c_{S'}(y)) \in S'$, $x <_{\mathbf{Q}} y$ such that $c_{S'}(x) = \left|S \cap \varphi^{-1}(x)\right| < |x|$, $c_{S'}(y) = \left|S \cap \varphi^{-1}(y)\right| > 0$ and x minimal, y maximal (in $\mathbf{Q}$) wrt those properties. Apply Definition 9.1, part (3), to x, y, decrementing $c_{S'}(y)$ to 0 or augmenting $c_{S'}(x)$ to $|x|$, to produce $S'' \in \mathfrak{I}(\overline{\mathbf{Q}}_{wk})$ with $\left|S''\right| = \left|S'\right| = |S|$ and $\Delta(S'') \geq \Delta(S') \geq \Delta(S)$. If $S'' \in \mathfrak{I}(\overline{\mathbf{Q}}_{st})$ then we are done, but if not we have eliminated at least one pair, x, y which prevented S' from being a member of $\mathfrak{I}(\overline{\mathbf{Q}}_{st})$. We may continue in the same manner until we do get $T \in \mathfrak{I}(\overline{\mathbf{Q}}_{st})$ with $|T| = |S|$ and $\Delta(T) \geq \Delta(S)$. □

9.1.2.2 Weak MWI-morphisms

To reverse the inequality of Lemma 9.1, part (2), and make it an equality, we would need to show that $\forall S \in \mathfrak{I}(\overline{\mathbf{Q}}_{wk})$, $\exists S' \in \mathfrak{I}(\mathbf{P})$ such that $\left|S'\right| = |S|$ and $\Delta(S') \geq \Delta(S)$. If this is the case, we call $\varphi : \mathbf{P} \to \overline{\mathbf{Q}}_{wk}$ a *weak MWI-morphism*. To make the definition effective, we offer the following (conditions (1) and (2) are again repeating the necessary conditions of Section 9.1.1)

Definition 9.2 *Let* $\mathbf{P}$ *be a weighted poset,* $\mathbf{Q}$ *a poset and* $\varphi : P \to Q$ *a function. Then* φ *is a Macaulay MWI-morphism,* $\varphi : \mathbf{P} \to \overline{\mathbf{Q}}_{wk}$, *if*

(1) φ *is order-preserving:* $\forall x, y \in \mathbf{P}$, $x \leq y$ *implies* $\varphi(x) \leq \varphi(y)$);
(2) φ *is weight-preserving:* $\forall x \in \mathbf{Q}$ *and* $0 \leq i \leq |x|$,

$$\Delta_{\overline{\mathbf{Q}}}(x, \{1, ..., i\}) = MWI\left(\varphi^{-1}(x), \Delta; i\right)$$

where $\varphi^{-1}(x)$ *inherits its partial order, cardinality and weight from* $\mathbf{P}$;
(3) $\forall y \in Q$, *the MWI problem on* $\varphi^{-1}(y)$ *has nested solutions* $S_{y,0} \subset S_{y,1} \subset ... \subset S_{y,|y|}$, *with* $\left|S_{y,j}\right| = j$, *which are also solutions of the MinShadow* $(x, y; j)$ *problem for every* $x, y \in Q$ *and such that* $S_{x,i} = \varphi^{-1}(x) \cap \overleftarrow{S_{y,j}}$ *when* $i = MinShadow(x, y; j)$.

Theorem 9.2 *A Macaulay MWI-morphism is a weak MWI-morphism.*

Proof If $S \in \mathfrak{I}(\overline{\mathbf{Q}}_{wk})$ and $c(x) = \max\{i : (x, i) \in I\}$ then

$$S' = \bigcup_{x \in \mathbf{Q}} S_{x,c(x)} \in \mathfrak{I}(\mathbf{P}),$$

$\left|S'\right| = |S|$ and $\Delta(S') = \Delta(S)$. □

Condition (3) in the definition of Macaulay MWI-morphism is so exacting that it might seem vacuous but it does have exemplars in the literature. The stability order of a finite Boolean lattice, with elements weighted by their marginal contribution to the shadow of a (any) stable set, makes the rank function a Macaulay MWI-morphism. In general, however, the condition is an unlikely one and difficult to verify even if true.

Definition 9.3 *$\varphi : \mathbf{P} \to \overline{\mathbf{Q}}_{in}$, where $\leq_{\overline{\mathbf{Q}}_{wk}} \subseteq \leq_{\overline{\mathbf{Q}}_{in}} \subseteq \leq_{\overline{\mathbf{Q}}_{st}}$, is an (intermediate) MWI-morphism if*

(1) *φ is order-preserving: $\forall x, y \in \mathbf{P}$, $x \leq y$ implies $\varphi(x) \leq \varphi(y)$);*
(2) *φ is weight-preserving: $\forall x \in \mathbf{Q}$ and $0 \leq i \leq |x|$,*

$$\Delta_{\overline{\mathbf{Q}}}(x, \{1, ..., i\}) = MWI\left(\varphi^{-1}(x), \Delta; i\right);$$

(3) *where $\varphi^{-1}(x)$ inherits its partial order, cardinality and weight from* **P**,
   (a) *$\forall S \in \Im(\mathbf{P}), \exists S' \in \Im\left(\overline{\mathbf{Q}}_{in}\right)$ such that $|S'| = |S|$ and $\Delta(S') \geq \Delta(S)$, and*
   (b) *$\forall S \in \Im\left(\overline{\mathbf{Q}}_{in}\right), \exists S' \in \Im(\mathbf{P})$ such that $|S'| = |S|$ and $\Delta(S') \geq \Delta(S)$.*

Theorem 9.3 *(The Fundamental lemma) If $\varphi : \mathbf{P} \to \overline{\mathbf{Q}}_{in}$ is an MWI-morphism then $\forall k \in \mathbb{Z}_+$,*

$$MWI(\mathbf{P}, \Delta; k) = MWI\left(\overline{\mathbf{Q}}_{in}, \Delta; k\right),$$

i.e. the MWI problem on **P** *is equivalent to that on $\overline{\mathbf{Q}}_{in}$.*

9.2 Examples

Our examples have been chosen small enough so that all calculations may be done by hand but large enough that the MWI-morphisms are not totally trivial.

9.2.1 EIP on the dodecahedron

Fig. 9.1 is adapted from Fig. 5.2. Each letter, x, labels a vertex of the dodecahedron and the adjacent number is its weight, $\Delta(x)$. With $\mathcal{Q} = \{A < B < C < D < E < F\}$, the following table represents φ :

X	$\varphi^{-1}(X)$	$\Delta(X)$
A	$\{a, b, c, d, f\}$	$(0, 1, 1, 1, 2)$
B	$\{e, g, h\}$	$(1, 1, 2)$
C	$\{i, j\}$	$(1, 2)$
D	$\{k, l\}$	$(1, 2)$
E	$\{m, n, p\}$	$(1, 2, 2)$
F	$\{o, q, r, s, t\}$	$(1, 2, 2, 2, 3)$

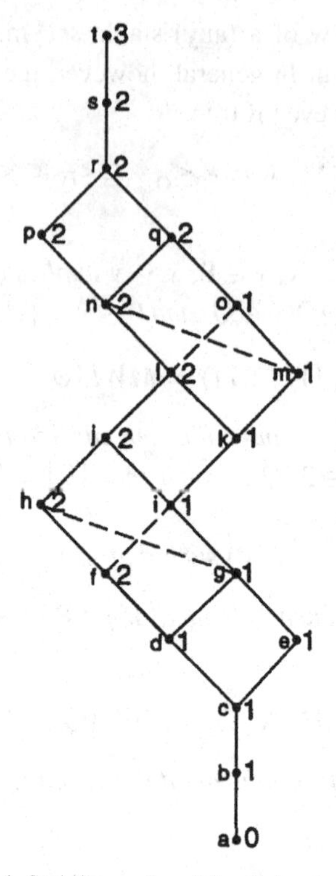

Fig. 9.1 Stability order of the dodecahedron.

To facilitate the verification of Definition 9.1, part (3), note that $\forall X$, $\varphi^{-1}(X)$ has a maximum element and a minimum element (in this case $\varphi^{-1}(X)$ happens to be totally ordered). The third column gives the list of weights for X.

The following table represents the *MinShadow* function for φ:

X	Y	k	$MinShadow(X, Y; k)$
A	B	1	3
A	B	2	4
B	C	1	2
B	D	1	2
B	E	1	2
C	D	1	2
C	E	1	1
D	E	1	1
E	F	1	1
E	F	2	2

Lastly we have the table which for each $X < Y$, j and $i \geq MinShadow(X, Y; j)$ gives $M = MWI(X; i) + MWI(Y; j)$ and $N = MWI(X; i + j)$ or $O = \Delta(X) + MWI(Y; i + j - |X|)$.

X	Y	j	i	M	N	O
A	B	1	3	$2+1=3$	3	
A	B	1	4	$3+1=4$	5	
A	B	2	4	$3+2=5$		$5+1=6$
B	C	1	2	$2+1=3$	4	
B	D	1	2	$2+1=3$	4	
B	E	1	2	$2+1=3$	4	
C	D	1	1	$1+1=2$	3	
C	E	1	1	$1+1=2$	3	
D	E	1	1	$1+1=2$	3	
E	F	1	1	$1+1=2$	3	
E	F	1	2	$1+3=4$	5	
E	F	2	2	$3+3=6$		$5+1=6$

Since the value appearing in the rightmost column is always at least that in the previous column, we have shown that φ is a skeletal MWI-morphism. From that, and the fact that $\mathcal{Q}$ and $\varphi^{-1}(X)$ are totally ordered $\forall X \in \mathcal{Q}$, we conclude that the dodecahedron has nested solutions with respect to the ordering

$$(a, b, c, d, f, e, g, h, i, j, k, l, m, n, p, o, q, r, s, t)$$

of its vertices. This was previously demonstrated in Chapter 5, Section 5.1.1, by generating the derived network of the stability order.

9.2.2 EIP on BS_4

Computer scientists call the Cayley graph of the symmetric group, S_n, with respect to the consecutive transpositions, $\{(i, i+1) : 1 \leq i < n\}$, *the bubble-sort graph*. BS_n is a Coxeter group, so the theory of stabilization applies to its edge-isoperimetric problem (see Chapter 5). Another way of looking at BS_n is as the graph of the *permutohedron*, the convex polytope generated by the set

$$\{(\pi(1), \pi(2), ..., \pi(n)) : \pi \in S_n\}.$$

Because $\forall \pi \in S_n$, $\sum_{i=1}^{n} \pi(i) = \sum_{j=1}^{n} j = \binom{n+1}{2}$, the permutohedron is only $(n-1)$-dimensional. In 1911 already, Schoute had noted that the

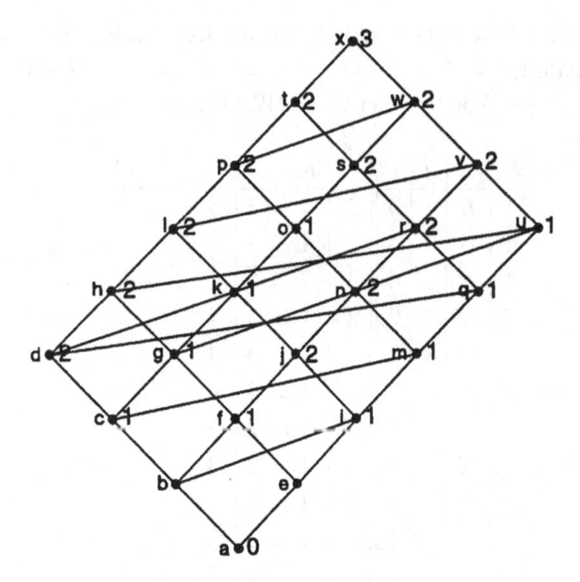

Fig. 9.2 Stability order of BS_4.

three-dimensional permutohedron is isomorphic to the snub-octahedron (the octahedron with each vertex sliced off to make a square face, see [**93**], pp. 17, 18) which means that its symmetry group is that of the octahedron (and cube). All the reflections of that larger (order 48 rather than $4! = 24$) Coxeter group are stabilizing for BS_4 and the resulting stability order (with weights for EIP) is shown in Fig. 9.2.

The letters represent 4-permutations according to the following table

a	1234	i	3124	q	3142
b	2134	j	3214	r	4132
c	1243	k	4123	s	3241
d	2143	l	4213	t	4231
e	1324	m	1342	u	3412
f	2314	n	1432	v	4312
g	1423	o	2341	w	3421
h	2413	p	2431	x	4321

With $\mathcal{Q} = \{A < B < C < D < E < F\}$, the following table represents φ:

X	$\varphi^{-1}(X)$	$\Delta(X)$
A	$\{a, b, c, d\}$	$(0, 1, 1, 2)$
B	$\{e, f, g, h\}$	$(1, 1, 1, 2)$
C	$\{i, j, k, l\}$	$(1, 2, 1, 2)$
D	$\{m, n, o, p\}$	$(1, 2, 1, 2)$
E	$\{q, r, s, t\}$	$(1, 2, 2, 2)$
F	$\{u, v, w, x\}$	$(1, 2, 2, 3)$

Exercise 9.1 *Verify that this φ gives a skeletal MWI-morphism.*

This proves that BS_4 has nested solutions for EIP and that the optimal total extension is given by alphabetic order.

9.2.3 EIP on the 24-cell

The stability order of the (graph of) the 24-cell (see [**28**]) is given in Fig. 9.3 (from Chapter 5 where it is Fig. 5.9).

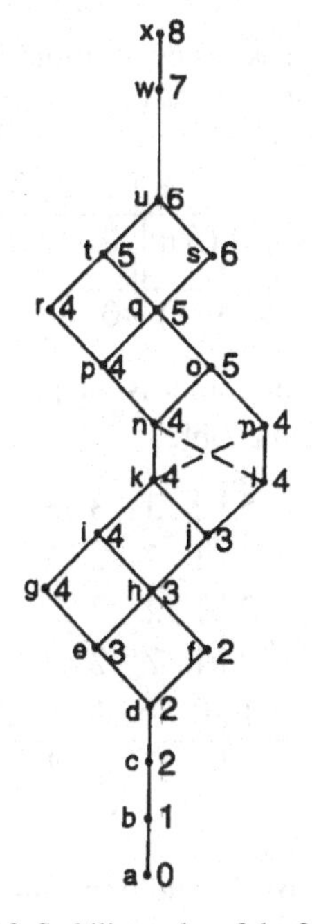

Fig. 9.3 Stability order of the 24-cell.

With $Q = \{A < B < C < D < E < F < G\}$, the following table represents φ:

X	$\varphi^{-1}(X)$	$\Delta(X)$
A	$\{a, b, c, d, e, g\}$	$(0, 1, 2, 2, 3, 4)$
B	$\{f, h, i\}$	$(2, 3, 4)$
C	$\{j, k\}$	$(3, 4)$
D	$\{l, m\}$	$(4, 4)$
E	$\{n, o\}$	$(4, 5)$
F	$\{p, q, s\}$	$(4, 5, 6)$
G	$\{r, t, u, v, w, x\}$	$(4, 5, 6, 6, 7, 8)$

Exercise 9.2 *Verify that this φ gives a skeletal MWI-morphism, proving that (the graph of) the 24-cell has nested solutions and that the optimal total extension is*

$$a, b, c, d, e, g, f, h, i, j, k, l, m, n, o, p, q, s, r, t, u, v, w, x.$$

9.2.4 $\mathbb{Z}_5 \times \mathbb{Z}_5$

$\mathbb{Z}_5$, the graph of the pentagon, has nested solutions for the EIP and the maximum values for $|I(S)|$ and the resulting Δ-sequence are given by the table

k	1	2	3	4	5
$\max_{S \subseteq \mathbb{Z}_5, \|S\|=k} \|I(S)\|$	0	1	2	3	5
$\Delta(k)$	0	1	1	1	2

For $\mathbb{Z}_5 \times \mathbb{Z}_5$, the compressibility order is the product of total orders, $\mathcal{T}_5 \times \mathcal{T}_5$, weights being summarized in the table

i_2						
	5	2	3	3	3	4
	4	1	2	2	2	3
i_2	3	1	2	2	2	3
	2	1	2	2	2	3
	1	0	1	1	1	2
	Δ	1	2	3	4	5
				i_1		

(see Lemma 6.3 of Chapter 6). In addition, interchanging i_1 and i_2 gives a reflective symmetry which adds the relations $(i_1, i_2) < (i_2, i_1)$ if $i_1 < i_2$. If we assume that $\mathcal{T}_5 \times \mathcal{T}_5$ also has nested solutions, then we can find the optimal total extension of it by locally maximizing the weight of successive augmentations. This leads us to the numbering

	5	9	10	15	24	25
	4	7	8	14	22	23
i_2	3	5	6	13	20	21
	2	2	4	12	17	19
	1	1	3	11	16	18
	η	1	2	3	4	5
				i_1		

To prove that the initial segments,

$$S_k(\eta) = \{(i_1, i_2) \in \mathcal{T}_5 \times \mathcal{T}_5 : \eta(i_1, i_2) \leq k\}, \ 0 \leq k \leq 25,$$

of this numbering are optimal, we define an MWI-morphism with $Q = \{A < B < C\}$ and $\varphi : \mathbb{Z}_5 \times \mathbb{Z}_5 \to Q$ given by

X	$\varphi^{-1}(X)$	$\Delta(X)$
A	$\{1\} \times \mathcal{T}_5 + \{2\} \times \mathcal{T}_5$	$(0, 1, 1, 2, 1, 2, 1, 2, 2, 3)$
B	$\{3\} \times \mathcal{T}_5$	$(1, 2, 2, 2, 3)$
C	$\{4\} \times \mathcal{T}_5 + \{5\} \times \mathcal{T}_5$	$(1, 2, 2, 3, 2, 3, 2, 3, 3, 4)$

Verifying that this φ does determine a strong MWI-morphism involves a couple of extra steps at this point, compared to previous examples. $\varphi^{-1}(A)$ and $\varphi^{-1}(C)$ are not totally ordered so we must solve the MWI problem on each of them. That may be accomplished by another MWI-morphism $\varphi' : \varphi^{-1}(A) \to \{1 < 2 < 3 < 4 < 5\}$ defined by $\varphi'((i_1, i_2)) = i_2$. There is a small hitch in verifying Definition 9.1 for φ'. The same problem comes up later so let us examine it in this transparent example: φ' is not a skeletal *MWI*-morphism because Definition 9.1, part (3a) fails for $x = 1, y = 5$ and $i = j = 1$. The ideals of $\mathcal{T}_5 \times \mathcal{T}_5$ containing $(1, 5)$ as a maximal element and not containing $(2, 1)$ cannot be reduced as in the proof of Theorem 9.1. However, there is only one such ideal, $\{1\} \times \mathcal{T}_5$. $|\{1\} \times \mathcal{T}_5| = 5$ and $\Delta(\{1\} \times \mathcal{T}_5) = 5$ whereas $|S_5(\eta)| = 5$ and $\Delta(S_5(\eta)) = 5$ also. Thus with $S' = S_5(\eta)$ we complete the proof that φ' is a strong MWI-morphism. This proves that $\varphi^{-1}(A)$ has nested solutions and that the optimal numbering is η restricted to $\varphi^{-1}(A)$. The weights on $\varphi^{-1}(C)$ only differ from those on $\varphi^{-1}(A)$ by a constant so we get the same result there too.

Problem 9.1 *Show that the EIP on products of $\mathbb{Z}_n$'s, $n \leq 4$, have lexicographic nested solutions.*

Problem 9.2 *Solve the EIP on $\mathbb{Z}_6 \times \mathbb{Z}_6$, showing that it does not have nested solutions.*

Problem 9.3 *Solve the EIP on $V_{20} \times V_{20}$. Does it have nested solutions?*

9.3 Application I: The pairwise product of Petersen graphs

Compression can give solutions of the EIP on graphs $G = G_1 \times G_2 \times ... \times G_d$, of unlimited size but is not applicable to irreducible graphs ($d = 1$) and not very effective when $d = 2$. If G is irreducible but highly symmetric, or if $d = 2$ and $G_1 = G_2$, stabilization can be effective, but there are many interesting regular graphs for which neither compression or stabilization nor the two together are enough to achieve a solution. Those are the cases for which MWI-morphisms were made.

9.3.1 The product of Petersen graphs

A diagram of the Petersen graph, P, is shown in Fig. 6.1 of Chapter 6 with its optimal numbering. Following that, in Section 6.4.3, is the weighted compressibility order, $\mathcal{T}_{10} \times \mathcal{T}_{10}$ (repeated here for convenience)

	10	3	4	4	4	5	4	5	5	5	6
	9	2	3	3	3	4	3	4	4	4	5
	8	2	3	3	3	4	3	4	4	4	5
	7	2	3	3	3	4	3	4	4	4	5
	6	1	2	2	2	3	2	3	3	3	4
i_2	5	2	3	3	3	4	3	4	4	4	5
	4	1	2	2	2	3	2	3	3	3	4
	3	1	2	2	2	3	2	3	3	3	4
	2	1	2	2	2	3	2	3	3	3	4
	1	0	1	1	1	2	1	2	2	2	3
	Δ	1	2	3	4	5	6	7	8	9	10
						i_1					

Assuming that $\mathcal{T}_{10} \times \mathcal{T}_{10}$ has nested solutions, Bezrukov, Das and Elsässer found the optimal numbering by optimizing locally:

The Bezrukov–Das–Elsässer numbering of $\mathcal{T}_{10} \times \mathcal{T}_{10}$

i_2											
	10	19	20	30	49	50	69	70	80	99	100
	9	17	18	29	47	48	67	68	79	97	98
	8	15	16	28	45	46	65	66	78	95	96
	7	13	14	27	43	44	63	64	77	93	94
	6	11	12	26	41	42	61	62	76	91	92
i_2	5	9	10	25	39	40	59	60	75	89	90
	4	7	8	24	37	38	57	58	74	87	88
	3	5	6	23	35	36	55	56	73	85	86
	2	3	4	22	33	34	53	54	72	83	84
	1	1	2	21	31	32	51	52	71	81	82
	$\mathcal{P}_2$	1	2	3	4	5	6	7	8	9	10
						i_1					

They proved the optimality of their numbering, $\mathcal{P}_2$, by brute force, calculating $|E(S)|$ for all $\binom{20}{10} = 184\,756$ ideals. Before the introduction of MWI-morphisms, a humanly verifiable proof seemed impossible.

Looking at $\mathcal{P}_2$, it seems natural to define

$$\mathcal{Q} = \{A < B < C < D < E < F\}$$

and $\varphi : \mathcal{T}_{10} \times \mathcal{T}_{10} \to \mathcal{Q}$ by

X	$\varphi^{-1}(X)$	$\Delta(X)$
A	$\{1\} \times \mathcal{T}_{10} + \{2\} \times \mathcal{T}_{10}$	$(0, 1, 1, 2, 1, 2, 1, 2, 2, 3, 1, 2, 2, 3, 2, 3, 2, 3, 3, 4)$
B	$\{3\} \times \mathcal{T}_{10}$	$(1, 2, 2, 2, 3, 2, 3, 3, 3, 4)$
C	$\{4\} \times \mathcal{T}_{10} + \{5\} \times \mathcal{T}_{10}$	$(1, 2, 2, 3, 2, 3, 2, 3, 3, 4, 2, 3, 3, 4, 3, 4, 3, 4, 4, 5)$
D	$\{6\} \times \mathcal{T}_{10} + \{7\} \times \mathcal{T}_{10}$	$(1, 2, 2, 3, 2, 3, 2, 3, 3, 4, 2, 3, 3, 4, 3, 4, 3, 4, 4, 5)$
E	$\{8\} \times \mathcal{T}_{10}$	$(2, 3, 3, 3, 4, 3, 4, 4, 4, 5)$
F	$\{9\} \times \mathcal{T}_{10} + \{10\} \times \mathcal{T}_{10}$	$(2, 3, 3, 4, 3, 4, 3, 4, 4, 5, 3, 4, 4, 5, 4, 5, 4, 5, 5, 6)$

$\mathcal{P}_2$, restricted to the first two columns, $\{1\} \times \mathcal{T}_{10} + \{2\} \times \mathcal{T}_{10}$, may be shown optimal by another MWI-morphism, $\varphi' : \{1\} \times \mathcal{T}_{10} + \{2\} \times \mathcal{T}_{10} \to \mathcal{Q}' = \{A' < B'\}$ defined by

X'	$\varphi'^{-1}(X')$	$\Delta(X')$
A'	$\mathcal{T}_2 \times \mathcal{T}_5$	$(0, 1, 1, 2, 1, 2, 1, 2, 2, 3)$
B'	$\mathcal{T}_2 \times (\mathcal{T}_{10} - \mathcal{T}_5)$	$(1, 2, 2, 3, 2, 3, 2, 3, 3, 4)$

the list-weights having been already calculated in the $\mathcal{T}_5 \times \mathcal{T}_5$ example. There are 25 cases to consider; all but two satisfy Definition 9.1, part (3), and those are easily rectified as in the $\mathcal{T}_5 \times \mathcal{T}_5$ example. The restrictions of η to the other pairs of columns, $\varphi^{-1}(C)$, $\varphi^{-1}(D)$ and $\varphi^{-1}(F)$, are also optimal since their list-weights differ from that on $\varphi^{-1}(A)$ by a constant.

Then the *MinShadow* function, whose nontrivial domain is of cardinality 184, must be calculated, and the really tedious part by hand, the calculation of

$$MWI(X;i) + MWI(Y;j)$$

for all $X < Y$, j and $i \geq MinShadow(X,Y;j)$ for comparison with $MWI(X;i+j)$ or $\Delta(X) + MWI(Y;i+j-|X|)$. There are about 1200 such calculations. All but 41 satisfy Definition 9.1, part (3). Of those that fail, most involve zero or one ideal, as in the $\mathcal{T}_5 \times \mathcal{T}_5$ example. The only ones that involve more have $X - B$, $Y = F$, $j = 4$ and $3 \leq i \leq 5$. Those ideals all contain $\mathcal{T}_2 \times \mathcal{T}_{10} \cup \mathcal{T}_{10} \times \mathcal{T}_2$ which has $|\mathcal{T}_2 \times \mathcal{T}_{10} \cup \mathcal{T}_{10} \times \mathcal{T}_2| = 36$ and $\Delta(\mathcal{T}_2 \times \mathcal{T}_{10} \cup \mathcal{T}_{10} \times \mathcal{T}_2) = 76$. Their additional elements are an ideal in $(\mathcal{T}_8 - \mathcal{T}_2) \times (\mathcal{T}_5 - \mathcal{T}_2)$, a 6×3 rectangle which has $\binom{9}{6} = 84$ ideals, still nontrivial. If we show that $76 + MWI((\mathcal{T}_8 - \mathcal{T}_2) \times (\mathcal{T}_5 - \mathcal{T}_2), |\cdot|, \Delta; k) \leq \Delta(S_{36+k}(\eta))$, $1 \leq k \leq 18$, then we are done ($S' = S_{36+k}(\eta)$). This may be accomplished with yet another MWI-morphism, $\varphi'' : (\mathcal{T}_8 - \mathcal{T}_2) \times (\mathcal{T}_5 - \mathcal{T}_2) \to Q = \{A'' < B'' < C''\}$ defined by

X''	$\varphi''^{-1}(X'')$	$\Delta(X'')$
A''	$\{3\} \times (\mathcal{T}_5 - \mathcal{T}_2)$	$(2, 2, 3)$
B''	$\{4, 5\} \times (\mathcal{T}_5 - \mathcal{T}_2)$	$(, 2, 3, 2, 3, 3, 4)$
C''	$\{6, 7, 8\} \times (\mathcal{T}_5 - \mathcal{T}_2)$	$(2, 3, 3, 2, 3, 3, 4, 3, 4)$

which is even skeletal.

This proof of the optimality of the Bezrukov–Elsässer numbering involves about 1600 steps but is better than the brute force method by a factor of more than 100.

9.4 Application II: The EIP on the 600-vertex

9.4.1 How to repair broken inequalities

Basically, MWI-morphisms represent a divide-and-conquer method for solving an MWI problem, the partition $\{\varphi^{-1}(x) : x \in \mathbf{Q}\}$ giving the division of $\mathbf{P}$ into subposets representing subproblems and the quotient a simplified version of the original problem. But not every such partition will work (in fact very few will), so one must know how to divide in order to conquer. The fundamental intuition behind the definition of a strong MWI-morphism is that the blocks of the partition must have relatively high marginal weight so that for every cardinality there will be a solution which is essentially a union of those blocks. Definition 9.1 just makes that intuition quantitatively precise.

Finding the partitions for the examples of Section 9.2 and the application of Section 9.3 required a bit of fiddling which usually began by assuming that the problem had nested solutions and then generating the total order for it by maximizing the marginal weight of successive elements (starting with the null set). One can then examine that total order and break it up into promising segments. A good breakpoint is usually preceded by elements of relatively large weight and followed by those of relatively small weight. The test for a partition is the inequalities of Definition 9.1, part (3). If some pair of pairs, $(x, i), (y, j) \in \overline{Q}$ with $x <_{\mathbf{Q}} y$ and $i \geq MinShadow(x, y; j)$, does not satisfy the inequality (in which case we call it *broken* and $((x, i), (y, j))$ a *breaking pair*), the partition may still be saved by showing that it does satisfy the more general conditions for a strong MWI-morphism.

Definition 9.4 *For $\varphi : \mathbf{P} \to \mathbf{Q}$, order and weight preserving, let $BP(\varphi)$ be the set of all breaking pairs,$((x, i), (y, j))$ in $\overline{Q}$. Also, let*

$$X(\varphi) = \{x : \exists ((x, i), (y, j)) \in BP(\varphi)\},$$
$$Y(\varphi) = \{y : \exists ((x, i), (y, j)) \in BP(\varphi)\}.$$

For $x \in X(\varphi)$ let

$$BP(\varphi, x) = \{y \in Y(\varphi) : \exists ((x, i), (y, j)) \in BP(\varphi)\},$$

and for $y \in Y(\varphi)$ let

$$BP(\varphi, y) = \{x \in X(\varphi) : \exists ((x, i), (y, j)) \in BP(\varphi)\}.$$

The examples and applications of Sections 9.2 and 9.3 show how to cope with broken inequalities in an ad hoc way but as more of them appear, we need to be more systematic. Suppose that we have an ideal $S \in \mathfrak{I}(\mathbf{P})$ and iteratively apply the process in the proof of Theorem 9.1 to $\overline{\varphi}(S) \in \mathfrak{I}\left(\overline{\mathbf{Q}}_{wk}\right)$. At any step we have $S' \in \mathfrak{I}\left(\overline{\mathbf{Q}}_{wk}\right)$. If $\exists x <_{\mathbf{Q}} y$ such that $c_{S'}(x) = \max\left\{i : (x, i) \in S'\right\} < |x|$, x minimal with respect to that property, and $c_{S'}(y) > 0$, y maximal with respect to that property, for which the inequalities of Definition 9.1 (3) hold, then we say that S' is *reducible*. If $\forall x <_{\mathbf{Q}} y$ either $c_{S'}x = |x|$ or $c_{S'}(y) = 0$ (so $S' \in \mathfrak{I}\left(\overline{\mathbf{Q}}_{st}\right)$), then we say that S' is *completely reduced*. If $S' \in \mathfrak{I}\left(\overline{\mathbf{Q}}_{wk}\right)$ is irreducible but not completely reduced, then

$$A = \{x \text{ minimal in } \mathbf{Q} : \exists y \in \mathbf{Q} \text{ and } ((x, c_{S'}(x)), (y, c_{S'}(y))) \in BP(\varphi)\}$$

is an antichain in $X(\varphi)$ and

$$B = \{y \text{ maximal in } \mathbf{Q} : \exists x \in \mathbf{Q} \text{ and } ((x, c_{S'}(x)), (y, c_{S'}(y))) \in BP(\varphi)\}$$

is an antichain in $Y(\varphi)$. Furthermore, A and B are *interlocking,* in the sense that

(1) $\forall x \in A, \emptyset \neq \overrightarrow{x} \cap B \subseteq BP(x, \varphi)$ and
(2) $\forall y \in B, \emptyset \neq \overleftarrow{y} \cap A \subseteq BP(y, \varphi)$.

This follows directly from the definition of A, B and $BP(\varphi)$.

Definition 9.5 $\langle A, B\rangle = \{z \in Q : \exists x \in A, y \in B \text{ and } x \leq z \leq y\}$.

Theorem 9.4 *If $\forall A \subseteq X(\varphi), B \subseteq Y(\varphi)$, interlocking antichains, and $\forall i, j \in \mathbb{Z}_+$,*

$$\Delta\left(\overleftarrow{B} - \langle A, B\rangle\right) + MWI\left(\overline{\langle A, B\rangle}_{wk}, \Delta; i\right) + MWI\left(\left(\mathbf{Q} - \overrightarrow{A} - \overleftarrow{B}\right)_{st}, \Delta; j\right)$$

$$\leq MWI\left(\overline{\mathbf{Q}}_{st}, \Delta; \left|\overleftarrow{B} - \langle A, B\rangle\right| + i + j\right),$$

then $\varphi : \mathbf{P} \to \overline{\mathbf{Q}}_{st}$ is an MWI-morphism.

Proof Every $S \in \Im\left(\overline{\mathbf{Q}}_{wk}\right)$ which is irreducible but not completely reduced, determines a pair of interlocking antichains, $A \subseteq X(\varphi)$, $B \subseteq Y(\varphi)$. With respect to A and B, then the elements of S fall into three parts:

(1) a constant part, $S^0 = \overline{\overleftarrow{B} - \langle A, B\rangle} \subseteq S$, which is contained in all such ideals,
(2) $S' = S \cap \overline{\langle A, B\rangle} \in \Im\left(\overline{\langle A, B\rangle}_{wk}\right)$ and
(3) $S'' = S \cap \overline{\mathbf{Q} - \overrightarrow{A} - \overleftarrow{B}} \in \Im\left(\left(\mathbf{Q} - \overrightarrow{A} - \overleftarrow{B}\right)_{st}\right)$.

$\langle A, B\rangle$ and $\mathbf{Q} - \overrightarrow{A} - \overleftarrow{B}$ are independent, in the sense that $x \in \langle A, B\rangle$, $y \in \mathbf{Q} - \overrightarrow{A} - \overleftarrow{B}$ imply that x and y are incomparable($x \nleq y$ and $x \ngeq y$). Thus every $S \in \mathfrak{I}\left(\overline{\mathbf{Q}}_{wk}\right)$ is representable as $S = S^0 + S' + S''$ with $S' \in \mathfrak{I}\left(\overline{\langle A, B\rangle}_{wk}\right)$ and $S'' \in \mathfrak{I}\left(\left(\overline{\mathbf{Q} - \overrightarrow{A} - \overleftarrow{B}}\right)_{st}\right)$ being chosen independently. Also

$$\begin{aligned}
\Delta(S) &= \Delta\left(S^0\right) + \Delta\left(S'\right) + \Delta\left(S''\right) \\
&\leq \Delta\left(S^0\right) + MWI\left(\overline{\langle A, B\rangle}_{wk}, \Delta; \left|S'\right|\right) \\
&\quad + MWI\left(\left(\overline{\mathbf{Q} - \overrightarrow{A} - \overleftarrow{B}}\right)_{st}, \Delta; \left|S''\right|\right) \\
&\leq MWI\left(\overline{\mathbf{Q}}_{st}, \Delta; \left|S^0\right| + \left|S'\right| + \left|S''\right|\right) \\
&= \Delta\left(S'''\right) \text{ for some } S''' \in \mathfrak{I}\left(\overline{\mathbf{Q}}_{st}\right) \text{ with } \left|S'''\right| = |S|.
\end{aligned}$$

□

9.5 The calculation for V_{600}

In the previous sections we have laid out the theory upon which our solution of the EIP for the 600-vertex regular solid in four dimensions is based. In this section we describe the actual calculation, step by step, explaining how certain choices were made, showing intermediate results and explaining some small deviations from the theory (taken from [**52**]). The calculation consists of three major steps:

(1) Reduction of the EIP to an MWI problem by stabilization.
(2) Calculation of the quotient and verification of an MWI-morphism.
(3) Solution of the MWI problem on the quotient (range of the MWI-morphism).

These are described in the following three subsections.

9.5.1 Reducing the EIP

9.5.1.1 Representation

In order to do the calculation on a computer, we must have a representation of the 600-vertex regular solid which the computer can recognize and manipulate. One obvious possibility is to give it a list of vertices, 4-tuples of real numbers. Such a list is found in Section 8.7 of [**28**]. It consists of the permutations of $(\pm 2, \pm 2, 0, 0)$, $\left(\pm\sqrt{5}, \pm 1, \pm 1, \pm 1\right)$, $\left(\pm\tau, \pm\tau, \pm\tau, \pm\tau^{-2}\right)$, $\left(\pm\tau^2, \pm\tau^{-1}, \pm\tau^{-1}, \pm\tau^{-1}\right)$ along with the even permutations of $\left(\pm\tau^2, \pm\tau^{-2}, \pm 1, 0\right)$, $\left(\pm\sqrt{5}, \pm\tau^{-1}, \pm\tau, 0\right)$ and $\left(\pm 2, \pm 1, \pm\tau, \pm\tau^{-1}\right)$ where $\tau = \frac{1+\sqrt{5}}{2}$, the golden mean. Obviously $\sqrt{5}$ and τ are irrational and cannot be represented exactly by finite decimal expansions but we found that the standard five decimal places of accuracy was sufficient. We call this set V_{600}.

9.5.1.2 Fricke–Klein point

The Fricke–Klein point, p, may be any point in $\mathbb{R}^4$ which is not fixed by a symmetry of V_{600} (see [**48**]). However, we found it desirable to pick p so that it has a slightly stronger property, i.e. that the distances from p to members of V_{600} be distinct. The point we picked was

$$p = (1.1, -2,\ 2.8,\ 1.6)\,.$$

9.5.1.3 Reflective symmetries

A *reflection in* $\mathbb{R}^d$ is a linear transformation of the form

$$r_\lambda(x) = x - 2(x \cdot \lambda)\,\lambda,$$

where $\lambda \in \mathbb{R}^d$, $\|\lambda\| = 1$ and $x \cdot \lambda = \sum_{i=1}^{d} x_i \lambda_i$, the inner product of x and λ.

$$H_\lambda = \left\{x \in \mathbb{R}^d : x \cdot \lambda = 0\right\}$$

is the *fixed hyperplane of* r_λ. Since any reflective symmetry of V_{600} must map some $x \in V_{600}$ to $r_\lambda(x) = y \in V_{600}$, $x \neq y$, and then $\lambda = \pm\frac{y-x}{\|y-x\|}$, we can find all the reflective symmetries of V_{600} by taking all differences of x, $y \in V_{600}$ and normalizing. We just eliminate those which do not give symmetries. Actually, we used V_{120}, the dual of V_{600}, which has the same symmetry group (and whose representation is also given in Section 8.7 of [**28**]) since it had occurred in a previous calculation. In Section 12.6 of [**28**], Coxeter shows that there are 60 such reflective symmetries for V_{120} and V_{600}. We chose the orientation of λ_i so that $p \cdot \lambda_i < 0$ and then ordered the set

$$\Lambda = \{\lambda_1, \lambda_2, \ldots \lambda_{60}\}\,,$$

so that $-(p \cdot \lambda_i) < -(p \cdot \lambda_{i+1})$.

9.5.1.4 Basic reflections

Coxeter's theory of groups generated by reflections shows (see Chapter 5) that the fixed hyperplanes of reflection in Λ partition $\mathbb{R}^4$ into 14 400 congruent connected components called *chambers*, each a simplex with one face at infinity. The chamber which contains the Fricke–Klein point is called *fundamental*. The reflections whose fixed hyperplanes bound the fundamental chamber constitute a *basis* (minimal generating set) of the group. For finite Coxeter groups in $\mathbb{R}^d$, bases have cardinality d. Since $-(p \cdot \lambda_i)$ is just the distance from p to H_{λ_i}, λ_1 is obviously a basis element and the same holds for λ_2. However, λ_3 may not be basic. It will be, unless the perpendicular from p to H_{λ_3} passes through H_{λ_1} or H_{λ_2} before it gets to H_{λ_3}. In general this is equivalent to the equation

$$(p + t\lambda_i) \cdot \lambda_h = 0$$

having a solution $0 < t < -(p \cdot \lambda_i)$ for some h, $1 \leq h < i$.

Theorem 9.5 *λ_i is basic iff $\forall h < i$,*

$$\lambda_i \cdot \lambda_h \leq \frac{p \cdot \lambda_h}{p \cdot \lambda_i}.$$

Proof $0 = (p + t\lambda_i) \cdot \lambda_h = p \cdot \lambda_h + t(\lambda_i \cdot \lambda_h)$. If $\lambda_i \cdot \lambda_h = 0$ there is no solution. If $\lambda_i \cdot \lambda_h \neq 0$ the solution is

$$t = \frac{-(p \cdot \lambda_h)}{\lambda_i \cdot \lambda_h}$$

which is < 0 if $\lambda_i \cdot \lambda_h < 0$. If $\lambda_i \cdot \lambda_h > 0$,

$$\frac{-(p \cdot \lambda_h)}{\lambda_i \cdot \lambda_h} = t < -(p \cdot \lambda_i)$$

is equivalent to

$$\lambda_i \cdot \lambda_h > \frac{p \cdot \lambda_h}{p \cdot \lambda_i}$$

whose negation is

$$\lambda_i \cdot \lambda_h \leq \frac{p \cdot \lambda_h}{p \cdot \lambda_i}.$$

□

In Coxeter theory a basis for any Coxeter group is characterized by its pairwise inner products being represented by the Coxeter graph of the group. For the group, G_4, of symmetries of V_{120} and V_{600}, whose Coxeter graph is shown in Fig. 9.4 (see Table 5.1 in Chapter 5)

We used this to verify that the vectors we found above really are a basis, and then relabeled them $\{\lambda_1', \lambda_2', \lambda_3', \lambda_4'\}$.

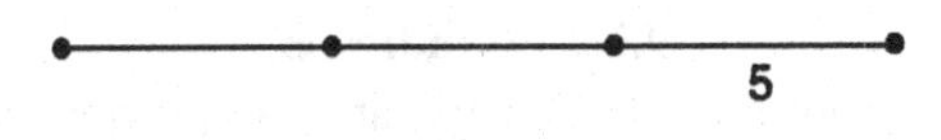

Fig. 9.4

9.5.1.5 The stability order

The Matsumoto–Verma theorem (see [**54**]) shows that the (weak and strong) stability orders on V_{600} are ranked and that the rank function is the same for both. This means that V_{600} is partitioned into ranks, $V_{600,r}$, $r = 0, 1, 2, \ldots$ and that members of $V_{600,r}$ are covered by members of $V_{600,r+1}$. $V_{600,0}$ consists of the single vertex contained in the closure of the fundamental chamber. It is also the vertex closest to p, the Fricke–Klein point. Given that we know $V_{600,r}$ for $r \geq 0$, then

$$V_{600,r+1} = \left\{r_{\lambda_i'}(x) : x \in V_{600,r},\ \lambda_i' \cdot x < 0,\ i = 1, 2, 3, \text{or } 4\right\}.$$

The covering relation in the (strong) stability order is then given by $x \lessdot y$ if, for some r, $x \in V_{600,r}$, $y \in V_{600,r+1}$ and $r_{\lambda_i}(x) = y$ for some i, $1 \leq i \leq 60$ (λ_i need not be basic but necessarily $\lambda_i \cdot x < 0$).

9.5.1.6 The weight, Δ

Exercise 6.7 of Chapter 6 notes that the EIP on the graph of any regular solid is equivalent to the MWI problem on its stability order, $\mathcal{S}$, with the weight function

$$\Delta(x) = |\{y \in V : \exists e \in E, \partial(e) = \{x, y\}\ \&\ y <_{\mathcal{S}} x\}|.$$

The edges of the solid generated by V_{600} are characterized by vertices at minimum distance

$$M = \min\{\|x - y\| > 0 : x, y \in V_{600}\}.$$

The Fricke–Klein order (**FK** the total order on V given by increasing distance from p) may be substituted for $\mathcal{S}$ in the definition of Δ since neighboring vertices are always comparable in $\mathcal{S}$ and **FK** is a total extension of $\mathcal{S}$. So for V_{600} we have

$$\Delta(x) = |\{y \in V_{600} :\ \|x - y\| = M\ \&\ y <_{\mathbf{FK}} x\}|.$$

The degree of any member of V_{600} is 4 so $0 \leq \Delta(x) \leq 4$ and since it is connected, 0 is achieved only by the initial, and 4 only by the terminal, vertex with respect to $\leq_{\mathbf{FK}}$.

9.5.2 Calculating the quotient, Q

9.5.2.1 Finding φ

We saw in the examples of Section 9.2 and the previous application, that finding the function $\varphi : \mathbf{P} \to \mathbf{Q}$ generally requires "a bit of fiddling". For the

600-vertex, however, there is another source of information: the solutions of the EIP for the other regular solids. These all suggest the hypothesis that for the 600-vertex we can find solutions which are unions of faces (dodecahedral cells). That pattern also holds for regular tessellations (see Chapter 7) which are infinite analogs of regular convex polytopes. This leads to the following definition for $\varphi : V_{600} \to V_{120}$. Recall that the elements of V_{120} and V_{600} are represented as points in $\mathbb{R}^4$, as given in Section 8.7 of [28]. Having chosen a Fricke–Klein point, p, the stability orders, $\mathcal{S}_{600}$ and $\mathcal{S}_{120}$ on V_{600} and V_{120} respectively, are determined. Then

$$\forall x \in V_{600},\ \varphi(x) = \min\{y \in \mathcal{S}_{120} : \|y - x\| = m\}$$

where

$$m = \min\left\{\left\|y' - x'\right\| : x' \in V_{600},\ y' \in V_{120}\right\}.$$

For a fixed $x \in V_{600}$ the elements of the set $\{y \in V_{120} : \|y - x\| = m\}$ are the vertices of the tetrahedral face centered on x. $\mathcal{S}_{120}$, restricted to that face, is the stability order of a tetrahedron and thus has a unique minimal element (see Chapter 5, Theorem 5.5) which is the designated $\varphi(x)$. For a fixed $y \in V_{120}$, $\varphi^{-1}(y) \subseteq \{x \in V_{600} : \|y - x\| = m\}$, a dodecahedral face. $\varphi^{-1}(y)$ is the subset of the vertices of that dodecahedral face which are not on lower (in $\mathcal{S}_{120}$) neighboring faces.

If we can verify that this function $\varphi : V_{600} \to V_{120}$ does give an MWI-morphism, then it will surely solve our problem. The stability order on V_{120} has 883 ideals. Since its width (maximum size of an antichain) is 4 (see [11]) and we can expect that the cardinality of $\varphi^{-1}(y)$ for most of those ys in 4-antichains will be near the average, $600/120 = 5$, the number of ideals in the quotient should be about $883\,(5)^4 < 10^6$, a number easily manageable by our 500 MHz PC. There may well be φs whose range has an even smaller number of ideals than the one we have chosen, but the effort required (in this case) to look for them would seem wasted. In other cases, however, there may be some point to it and we shall mention some of the possibilities for improvement in our comments at the end.

9.5.2.2 The MinShadow function

We calculated *MinShadow*$(x, y; j)$, for $x, y \in V_{120}$ and $1 \le j \le |y|$ with a variant of the program that was written to solve the MWI problem by brute force (see Section 9.5.3). It calculated the ideals of $\mathcal{S}_{600}$ which are generated by members of $\varphi^{-1}(y)$, in lexicographic order. For each such S and $\forall x \in V_{120}$, the cardinalities $\left|S \cap \varphi^{-1}(x)\right|$ were calculated and *MinShadow*$(x, y; j)$ updated. It was necessary to calculate the function for *all* pairs, not just those for which $x <_{\mathcal{S}_{120}} y$, because it turned out, to our surprise, that there are other pairs for which $MinShadow(x, y; |y|) > 0$. All such have $x <_{\mathbf{FK}} y$, so they induce a partial order on V_{120} which is stronger than $\mathcal{S}_{120}$ but still a suborder of **FK**. V_{120}, endowed with this induced partial order, is the quotient, **Q**.

9.5.2.3 The local MWI-function

$MWI\left(\varphi^{-1}(x), \Delta; i\right)$ essentially gives the weighting of $(x, i) \in \overline{Q}$ and completes the calculation of $\overline{\mathbf{Q}}_{st}$. We calculated it for all $x \in Q$ by the brute force method of Section 9.5.3. Having $MWI\left(\varphi^{-1}(x), \Delta; i\right)$ for all $x \in Q$, we then calculated $MWI\left(\overline{\mathbf{Q}}_{st}, \Delta; i\right)$ (which we hope will be $MWI(\mathcal{S}_{600}, \Delta; i)$ and is necessary for justifying the reduction by Theorem 9.4), also by brute force.

9.5.2.4 Checking inequalities

Having calculated the *MinShadow* and local MWI functions, verifying (or falsifying) the inequalities of Definition 9.1, part (3), was straightforward. The great majority (out of about 10^4) were validated but there were also 127 broken inequalities.

9.5.2.5 Repairing broken inequalities

We generated all pairs of interlocking antichains $A \subseteq X(\varphi)$, $B \subseteq Y(\varphi)$ by

(1) Generating all antichains $B \subseteq Y(\varphi)$. Since there is a one-to-one correspondence between ideals and antichains (the maximal elements of an ideal are an antichain which generates the original ideal), we used the routine for generating ideals, and for each ideal identified its maximal elements.
(2) Given B we found all antichains $A \subseteq \bigcup_{y \in B} BP(\varphi, y)$ satisfying the additional conditions.

For each such pair, A, B we verified the inequality of Theorem 9.4.

9.5.3 Solving the MWI problem on Q

9.5.3.1 Generating ideals

Given any total extension, **T** (we used the Fricke–Klein order), of the partial order **Q**, $\mathfrak{I}(\mathbf{Q})$ may be efficiently generated in lexicographic order. All we need to know about **Q** is the set of elements

$$cover(x) = \left\{y \in Q : x \lessdot_{\mathbf{Q}} y\right\},$$

which cover x in $\mathbf{Q}$. Let $S : \mathbf{Q} \to \{0, 1\}$ be the indicator function (i.e. an array wrt $\mathbf{T}$) of any ideal in $\mathbf{Q}$. Initialize it to all 0s, representing the empty ideal. If S is all 1s, representing the ideal Q, then we are done. If S is not all 1s then let $i_0 = \min\{i : S(i) = 0\}$. We generate S', by setting $S'(i) := S(i)$ for $i > i_0$, $S'(i_0) := 1$ and for $h = i_0 - 1$ down to 1 recursively setting

$$S'(h) := \begin{cases} 1 & \text{if } \exists j \in cov(h) \text{ such that } S'(j) = 1, \\ 0 & \text{otherwise.} \end{cases}$$

Lemma 9.2 *If $S \neq Q$ is an ideal in $\mathbf{Q}$ then S' is also an ideal, its successor in lexicographic order on the set of all ideals.*

9.5.3.2 Generating ideals of $\overline{\mathbf{Q}}_{st}$

Given an ideal, S, of $\mathbf{Q}$ we then identify the set, M, of maximal elements of S. The ideals $S' \in \mathfrak{I}\left(\overline{\mathbf{Q}}_{st}\right)$ for which

$$\left\{x \in \mathbf{Q} : \exists (x, i) \in S'\right\} = S$$

may then be generated by letting c run through all of its possible values, $0 < c(x) \leq |x|, \forall x \in M$, in lexicographic order. For each one we calculate $|S'|$ and $\Delta(S')$ and update $MWI(\mathcal{S}_{600}, \Delta; k)$ if necessary. This gave us the following solution of the MWI problem on $\mathcal{S}_{600}$:

k	1	2	3	4	5	6	7	8	9	10	11	12	13	14	15
$MWI(\mathcal{S}_{600}, \Delta; k)$	0	1	2	3	5	6	7	9	10	12	13	15	16	18	20

16	17	18	19	20	21	22	23	24	25	26	27	28	29	30	31	32
21	23	25	27	30	31	32	34	35	37	38	40	41	43	45	46	48

33	34	35	36	37	38	39	40	41	42	43	44	45	46	47	48	49
50	52	55	56	58	59	61	62	64	66	67	68	71	73	76	77	79

50	51	52	53	54	55	56	57	58	59	60	61	62	63	64	65
80	82	84	85	87	89	91	94	95	97	98	100	102	103	105	107

66	67	68	69	70	71	72	73	74	75	76	77	78	79
109	112	113	115	116	118	121	120	123	125	127	130	131	133

80	81	82	83	84	85	86	87	88	89	90	91	92	93
135	136	138	140	142	145	146	148	149	151	153	154	156	158

94	95	96	97	98	99	100	101	102	103	104	105	106	107
160	163	164	166	168	169	171	173	175	178	179	181	183	184

108	109	110	111	112	113	114	115	116	117	118	119	120	121
186	188	190	193	194	196	198	199	210	203	205	208	210	211

122	123	124	125	126	127	128	129	130	131	132	133	134	135
213	215	217	220	221	223	225	227	230	231	233	234	236	238

136	137	138	139	140	141	142	143	144	145	146	147	148	149
239	241	243	245	248	249	251	253	254	256	258	260	263	264

150	151	152	153	154	155	156	157	158	159	160	161	162	163
266	268	269	271	273	275	278	279	281	283	285	286	288	290

164	165	166	167	168	169	170	171	172	173	174	175	176	177
293	295	296	298	300	302	305	306	308	310	312	315	316	318

178	179	180	181	182	183	184	185	186	187	188	189	190	191
320	321	323	325	327	330	331	333	335	337	338	340	342	345

192	193	194	195	196	197	198	199	200	201	202	203	204	205
347	348	350	352	354	357	358	360	362	364	367	368	370	372

206	207	208	209	210	211	212	213	214	215	216	217	218	219
374	375	377	379	382	384	385	387	389	391	394	395	397	399

220	221	222	223	224	225	226	227	228	229	230	231	232	233
401	404	405	407	409	411	412	414	416	419	421	422	424	426

234	235	236	237	238	239	240	241	242	243	244	245	246	247
428	431	432	434	436	438	441	442	444	446	448	450	451	453

248	249	250	251	252	253	254	255	256	257	258	259	260	261
456	458	460	461	463	465	468	470	471	473	475	478	480	481

262	263	264	265	266	267	268	269	270	271	272	273	274	275
483	485	487	490	491	493	495	497	500	501	503	505	507	510

276	277	278	279	280	281	282	283	284	285	286	287	288	289
511	513	515	517	520	521	523	525	527	530	531	533	535	537

290	291	292	293	294	295	296	297	298	299	300	301	302	303
540	541	543	545	547	550	551	553	555	557	560	561	563	565

304	305	306	307	308	309	310	311	312	313	314	315	316	317
567	579	571	573	575	577	580	581	583	585	587	590	591	593

318	319	320	321	322	323	324	325	326	327	328	329	330	331
595	597	600	601	603	605	607	610	611	613	615	617	620	621

332	333	334	335	336	337	338	339	340	341	342	343	344	345
623	625	627	630	631	633	635	637	640	641	643	645	647	650

346	347	348	349	350	351	352	353	354	355	356	357	358	359
651	653	655	657	660	662	663	665	667	670	672	674	676	678

360	361	362	363	364	365	366	367	368	369	370	371	372	373
681	682	684	686	688	691	692	694	696	698	701	702	704	706

374	375	376	377	378	379	380	381	382	383	384	385	386	387
708	711	713	715	717	720	721	723	725	727	730	731	733	735

388	389	390	391	392	393	394	395	396	397	398	399	400	401
737	740	741	743	745	747	750	752	754	756	759	760	762	764

402	403	404	405	406	407	408	409	410	411	412	413	414	415
766	769	770	772	774	775	779	780	782	784	786	789	791	793

416	417	418	419	420	421	422	423	424	425	426	427	428	429
795	798	799	801	803	805	808	810	812	815	816	818	820	822

430	431	432	433	434	435	436	437	438	439	440	441	442	443
825	826	828	830	832	835	837	838	840	842	845	847	849	851

444	445	446	447	448	449	450	451	452	453	454	455	456	457
854	855	857	859	861	864	866	868	871	872	874	876	878	881

458	459	460	461	462	463	464	465	466	467	468	469	470	471
883	885	888	889	891	893	895	898	900	902	905	907	910	911

472	473	474	475	476	477	478	479	480	481	482	483	484	485
913	915	917	920	921	923	925	927	930	931	933	935	937	940

486	487	488	489	490	491	492	493	494	495	496	497	498	499
942	944	946	949	950	952	954	956	959	961	963	966	967	969

500	501	502	503	504	505	506	507	508	509	510	511	512
971	973	976	978	980	983	984	986	988	990	993	995	997

513	514	515	516	517	518	519	520	521	522	523
1000	1002	1005	1006	1008	1010	1012	1015	1017	1019	1022

524	525	526	527	528	529	530	531	532	533	534
1023	1025	1027	1029	1032	1034	1036	1039	1041	1044	1045

535	536	537	538	539	540	541	542	543	544	545
1047	1049	1051	1054	1056	1058	1061	1063	1066	1067	1069

546	547	548	549	550	561	552	553	564	555	556
1071	1073	1076	1078	1080	1083	1085	1088	1089	1091	1093

557	558	559	560	561	562	563	564	565	566	567
1095	1098	1100	1102	1105	1107	1110	1112	1115	1116	1118

568	569	570	571	572	573	574	575	576	577	578
1120	1122	1125	1127	1128	1132	1134	1137	1139	1142	1144

579	580	581	582	583	584	585	586	587	588	589
1147	1150	1151	1153	1155	1157	1160	1162	1164	1167	1169

590	591	592	593	594	595	596	597	598	599	600
1172	1174	1177	1179	1182	1185	1187	1190	1193	1196	1200

By the theory of Section 9.5.1 this is also the solution of the internal EIP which is equivalent to the external EIP on regular graphs. Thus the EIP on V_{600} has been solved.

9.6 Comments

The solutions of the EIP on the triangular and hexagonal tessellations of the plane given in Chapter 6 were based on reductions which can now be seen as MWI-morphisms.

The solution of the EIP on V_{600} is a capstone result, V_{600} being the last regular solid of any dimension for which the EIP had not been solved. A brute force solution would require the evaluation of all $2^{600} \simeq 4.15 \times 10^{180}$ subsets of vertices. This is more calculation than could be done in the lifetime of the universe by the fastest computer which could ever be built. Stabilization reduces that to about 10^{16} stable sets, still too many to calculate on any existing computer (we wasted 100 hours of CPU time because we underestimated the number of stable sets). With the MWI-morphism described in this chapter, there were less than 10^6 cases, which took about 30 seconds on our 500 MHz PC. Of course the program, about 50 pages of code and documentation, took several months to write.

The material of this chapter points up the need for better ways to estimate the number of ideals in a given poset, P. This problem is a generalization of the classical problem named after Dedekind, where the poset is a Boolean lattice on n generators. Unlike the Boolean lattice, however, where the obvious lower bound, $2^{A(P)}$, $A(P)$ being the maximum size of an antichain in P, is asymptotically sharp, $2^{A(P)}$ is often nowhere near $|\mathfrak{I}(P)|$

The definitions and applications of MWI-morphisms in this chapter were intended as a pilot project, demonstrating the feasibility and effectiveness of a divide-and-conquer strategy based on morphisms. This approach is unlikely to work with every hard problem, but the problems which arise in applications often have special structure which might be taken advantage of in this way. It is hoped that the methods, which solved the very beautiful but esoteric EIP on V_{600}, will be applicable to problems of wider interest.

Up to now, the only infinite families of graphs for which isoperimetric problems have been solved have been those generated by products of very simple irreducible graphs. In most cases, such as Lindsey's theorem (Theorem 6.4 of Chapter 6), the solution of the isoperimetric problem on the irreducible factors was trivial and the solution for products challenging but straightforward (with the theory of compression). We have now used up the stock of simple irreducible graphs with nested solutions for EIP or VIP, so if we are to exploit compression any further we must find more complex graphs which have nested solutions and whose pairwise products do also. Since, as we previously pointed out, compression is ineffectual on these problems, stabilization and MWI-morphisms are at present the only alternatives. We would even speculate that MWI-morphisms may provide solutions for infinite families of irreducible graphs and that they might not even have nested solutions!

Problem 9.4 Our solution of the EIP on BS_4 suggests that it can be solved for BS_n, $n > 4$. BS_5, however, does not have nested solutions (another application of Bezrukov's poset tools). Lubotzky ([**75**], problem 10.8.7) also mentions this as an interesting unsolved problem.

From early in the process of working up the material of this monograph, global methods were intended to be complementary to other algorithms and methods. For instance it is widely accepted that hill-climbing algorithms are ineffective when the domain has reflective symmetry, since that tends to create local maxima which are far from the maximum. Stabilization, by modding out reflective symmetry, may make hill-climbing more effective. That a successful global method can facilitate other methods is exemplified in our solution of the EIP on the 600-vertex where stabilization facilitates an MWI-morphism which in turn facilitates the brute computational power of the computer.

10

Passage to the limit

The failure of compression, or any combinatorial alternative, to solve the EIP on $P_{n_1} \times P_{n_2} \times ... \times P_{n_d}$ or the VIP on $K_{n_1} \times K_{n_2} \times ... \times K_{n_d}$, ultimately forced combinatorialists back to the old workhorse of classical mathematics: calculus. By this we mean passing to a continuous limit, solving the continuous problem by variational means and arguing that the essence of the combinatorial problem is preserved in the limit.

10.1 The Bollobas–Leader theorem

If we let $n_1 = n_2 = ... = n_d = n \to \infty$, then $\left(\frac{P_n}{n}\right)^d \to [0, 1]^d$, the d-dimensional unit cube. Given any ideal, $S \subseteq [0, 1]^d$, with the product partial order, then

$$\lim_{n\to\infty} \frac{\left|\left\{v \in (P_n)^d : \frac{v}{n^d} \in S\right\}\right|}{n^d} = \lambda(S),$$

the (d-dimensional) Lebesgue measure of S. Also, if $\sigma(S)$ is the area of the free boundary of S (that part of the boundary of S in the interior of $[0, 1]^d$), measured wrt the L_1 (taxicab) metric, then

$$\lim_{n\to\infty} \frac{\left|\Theta\left\{v \in (P_n)^d : \frac{v}{n^d} \in S\right\}\right|}{n^{d-1}} = \sigma(S).$$

Thus the analog of the EIP on $[0, 1]^d$ is, given $v \in [0, 1]$, to minimize $\sigma(S)$ over all ideals $S \subseteq [0, 1]^d$ such that $\lambda(S) = v$. We claim that the only critical (locally optimal) stable ideals for the EIP on $[0, 1]^d$ are $S(j, d, t) = [0, 1]^{d-j} \times [0, t]^j$, $1 \le j \le d$, and their dual-complements (see Section 1.2.2.3 of Chapter 1), $S^*(j, d, t) = [0, 1]^d - [1 - t, 1]^j \times [0, 1]^{d-j}$.

Since $v = \lambda(S(j, d, t)) = t^j$, $0 \le t \le 1$ and $\sigma(S(j, d, t)) = jt^{j-1}$, we have $\sigma_j(v) = jv^{1-\frac{1}{j}}$. Also, since $\lambda(S^*(j, d, t)) = 1 - \lambda(S(j, d, 1 - t))$ and

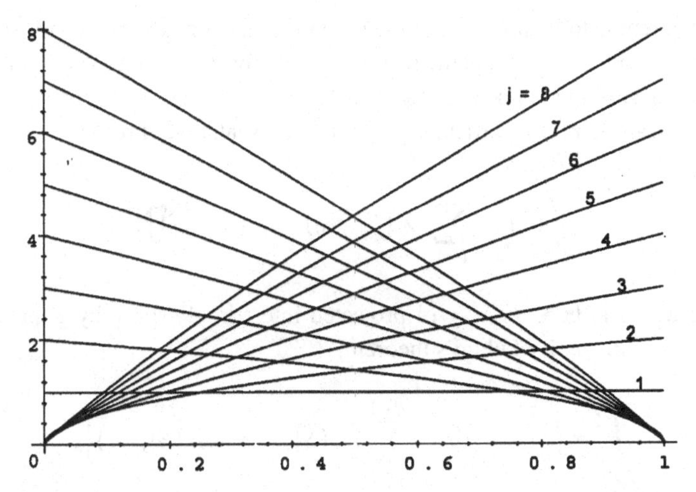

Fig. 10.1 Graphs of the Bollobas–Leader critical functions, $\sigma_j(v)$ and $\sigma_j^*(v)$.

$\sigma(S^*(j,d,t)) = \sigma(S(j,d,1-t))$, we have $\sigma_j^*(v) = j(1-v)^{1-\frac{1}{j}}$. The lower envelope of the graphs in Fig. 10.1 gives the function

$$F_d(v) = \min_{1\le j\le d}\left\{\sigma_j(v), \sigma_j^*(v)\right\}$$

$$= \begin{cases} \sigma_j(v) & \text{if } v_j \le v \le v_{j-1} \\ \sigma_d(v) & \text{if } 0 \le v \le v_d \\ \sigma_j(1-v) & \text{if } v_j \le 1-v \le v_{j-1} \\ \sigma_d(1-v) & \text{if } 0 \le 1-v \le v_d \end{cases}$$

where

$$v_j = \begin{cases} \frac{1}{2} & \text{if } j = 0 \\ \left(\frac{j}{j+1}\right)^{j(j+1)} & \text{if } j \ge 1. \end{cases}$$

Note that F_d, being the pointwise minimum of concave (downward) functions, is concave. Also, v_j is independent of n and $v_1 = \frac{1}{4}$ so $F_d(v) = \sigma_1(v) = 1$ if $\frac{1}{4} \le v \le \frac{3}{4}$. In addition, $v_j \simeq e^{-(j+1)}$ as $j \to \infty$.

Theorem 10.1 *(Bollobas and Leader* [**22**]*) For any $S \subseteq [0,1]^d$, $\sigma(S) \ge F_d(\lambda(S))$.*

Proof We may assume that S is one-dimensionally compressed and therefore an ideal in the product order on $[0,1]^d$. Also we may assume that S is stable.

The theorem is to be proved by induction on d and already proven for $d = 1$. Assume then that $d > 1$ and the theorem is true for $d - 1$. If $\lambda(S) = 0$ or 1 the result is trivial, so assume that $0 < \lambda(S) < 1$.

For ideals there is a particularly simple representation of $\sigma(S)$,

$$\sigma(S) = \sum_{i=1}^{d} \left[\lambda\left(S|_{x_i=0}\right) - \lambda\left(S|_{x_i=1}\right)\right],$$

where $S|_{x_i=c}$ is $\{x \in S : x_i = c\}$ projected into $(d-1)$-space by eliminating the ith coordinate. By Fubini's theorem,

$$\int_0^1 \sigma\left(S|_{x_d=t}\right) \mathrm{d}t = \sum_{i=1}^{d-1} \left[\lambda\left(S|_{x_i=0}\right) - \lambda\left(S|_{x_i=1}\right)\right],$$

so

$$\begin{aligned}
\sigma(S) &= \lambda\left(S|_{x_d=0}\right) - \lambda\left(S|_{x_d=1}\right) + \int_0^1 \sigma\left(S|_{x_d=t}\right) \mathrm{d}t \\
&\geq \lambda\left(S|_{x_d=0}\right) - \lambda\left(S|_{x_d=1}\right) + \int_0^1 F_{d-1}\left(\lambda\left(S|_{x_d=t}\right)\right) \mathrm{d}t
\end{aligned}$$

by the inductive hypothesis. Since S is an ideal,

$$\lambda\left(S|_{x_d=0}\right) \leq \lambda\left(S|_{x_d=t}\right) \leq \lambda\left(S|_{x_d=1}\right)$$

so $\exists \alpha(t)$, $0 \leq \alpha(t) \leq 1$, such that

$$\lambda\left(S|_{x_d=t}\right) = \alpha(t)\,\lambda\left(S|_{x_d=0}\right) + (1 - \alpha(t))\,\lambda\left(S|_{x_d=1}\right).$$

Since F_{d-1} is concave,

$$\begin{aligned}
&\int_0^1 F_{d-1}\left(\lambda\left(S|_{x_d=t}\right)\right) \mathrm{d}t \\
&\geq \int_0^1 \left[\alpha(t)\, F_{d-1}\left(\lambda\left(S|_{x_d=0}\right)\right) + (1 - \alpha(t))\, F_{d-1}\left(\lambda\left(S|_{x_d=1}\right)\right)\right] \mathrm{d}t \\
&= \alpha F_{d-1}\left(\lambda\left(S|_{x_d=0}\right)\right) + (1 - \alpha)\, F_{d-1}\left(\lambda\left(S|_{x_d=1}\right)\right),
\end{aligned}$$

where $\alpha = \int_0^1 \alpha(t)\,\mathrm{d}t$, so $0 \leq \alpha \leq 1$. If $\alpha = 0$ or 1, then $S|_{x_d=0} = S|_{x_d=1}$ and we are done. If not, we have

$$\begin{aligned}\sigma(S) &\geq \lambda\left(S|_{x_d=0}\right) - \lambda\left(S|_{x_d=1}\right) \\ &\quad + \alpha F_{d-1}\left(\lambda\left(S|_{x_d=0}\right)\right) \\ &\quad + (1-\alpha) F_{d-1}\left(\lambda\left(S|_{x_d=1}\right)\right) \\ &= \frac{\lambda(S) - \lambda\left(S|_{x_d=1}\right)}{\alpha} \\ &\quad + \alpha F_{d-1}\left(\lambda\left(S|_{x_d=1}\right) + \frac{\lambda(S) - \lambda\left(S|_{x_d=1}\right)}{\alpha}\right) \\ &\quad + (1-\alpha) F_{d-1}\left(\lambda\left(S|_{x_d=1}\right)\right).\end{aligned}$$

Define

$$H(x) = \frac{\lambda(S) - x}{\alpha} + \alpha F_{d-1}\left(x + \frac{\lambda(S) - x}{\alpha}\right) + (1-\alpha) F_{d-1}(x),$$

for

$$0 \leq x \leq \lambda(S) < 1 \text{ and } 0 \leq x + \frac{\lambda(S) - x}{\alpha} \leq 1.$$

The last inequality is equivalent to $\frac{\lambda(S)-\alpha}{1-\alpha} \leq x$ which is stronger than $0 \leq x$ if $\alpha < \lambda(S)$. Since H is the sum of three concave functions, it is itself concave and, since the minimum of a concave function on a finite interval is at an endpoint, it follows that

$$\begin{aligned}\sigma(S) &\geq H\left(\lambda\left(S|_{x_d=1}\right)\right) \\ &\geq \min \begin{cases} \left\{H\left(\frac{\lambda(S)-\alpha}{1-\alpha}\right), H(\lambda(S))\right\} & \text{if } \alpha \leq \lambda(S) \\ \{H(0), H(\lambda(S))\} & \text{if } \alpha > \lambda(S). \end{cases}\end{aligned}$$

Now

$$\begin{aligned} H(0) &= \frac{\lambda(S)}{\alpha} + \alpha F_{d-1}\left(\frac{\lambda(S)}{\alpha}\right) \\ H(\lambda(S)) &= F_{d-1}(\lambda(S)) \\ H\left(\frac{\lambda(S)-\alpha}{1-\alpha}\right) &= \frac{1-\lambda(S)}{1-\alpha} + (1-\alpha) F_{d-1}\left(\frac{\lambda(S)-\alpha}{1-\alpha}\right). \end{aligned}$$

Letting

$$v = \begin{cases} \frac{\lambda(S)}{\alpha} & \text{if } \alpha > \lambda(S) \\ \lambda(S) & \text{if } \alpha = \lambda(S) \\ \frac{\lambda(S)-\alpha}{1-\alpha} & \text{if } \alpha < \lambda(S) \end{cases}$$

we have that $v < \lambda(S)$ iff $\alpha > \lambda(S)$. If we define

$$h(v) = \begin{cases} v + \frac{\lambda(S)}{v} F_{d-1}(v) & \text{if } v < \lambda(S) \\ F_{d-1}(\lambda(S)) & \text{if } v = \lambda(S) \\ 1 - v + \frac{1-\lambda(S)}{1-v} F_{d-1}(v) & \text{if } v > \lambda(S). \end{cases}$$

then the preceding inequality becomes $\sigma(S) \geq \min h(v)$. The minimum is assumed at some point, v_0, in the interval $(0, 1)$ since h is continuous except at $v = \lambda(S)$ where $\lim_{v\to\lambda(S)} h(v) > h(\lambda(S))$ and $\lim_{v\to 0} h(v) = \infty = \lim_{v\to 1} h(v)$.

Given v, $0 \leq v \leq \frac{1}{2}$, let $S(d, v) = S(j, d, t)$, described above such that $\lambda(S(j, d, t)) = v$ and $\sigma(S(j, d, t)) = F_d(v)$. If there are two such sets, either one may be $S(j, d, t)$. For $\frac{1}{2} < v \leq 1$ let $S(d, v) = S^*(j, d, t)$. Then define an ideal

$$T(v) = \begin{cases} S(d-1, v) \times \left[0, \frac{\lambda(S)}{v}\right] & \text{if } v < \lambda(S) \\ S(d-1, v) \times [0, 1] & \text{if } v = \lambda(S) \\ (S(d-1, v) \times [0, 1]) \cup \left(\left[0, \frac{\lambda(S)-v}{1-v}\right] \times [0, 1]^{d-1}\right) & \text{if } v > \lambda(S) \end{cases}$$

and $\forall v, 0 < v < 1, \lambda(T(v)) = \lambda(S)$ and $\sigma(T(v)) = h(v)$.

To complete the proof, we need only show that the optimal ideal, $T = T(v_0)$ is of the form $S(j, d, t) = [0, 1]^j \times [0, t]^{d-j}$ or $S^*(j, d, t) = [0, 1]^d - [1-t, 1]^j \times [0, 1]^{d-j}$. There are four cases:

	$\lambda(S) < v_0$	$\lambda(S) > v_0$
$v_0 < \frac{1}{2}$	I	III
$v_0 > \frac{1}{2}$	II	IV

Since III is the dual-complement of II and IV is the dual-complement of I (see Section 1.2.2.3 of Chapter 1), we need only consider cases I and II. In case I, $T(v_0) = [0, 1]^j \times [0, t]^{d-1-j} \times [0, s]$, for some j. If $s \neq t$, then $[0, t] \times [0, s]$ may be replaced by $\left[0, (st)^{1/2}\right]^2$ contradicting the optimality of $T(v_0)$. Therefore $T(v_0) = [0, 1]^j \times [0, t]^{d-j}$. In case II, $T(v_0) = \left([0, 1]^d - [1-t, 1]^j \times [0, 1]^{d-j}\right) \times [0, s]$ which also contains a section $[0, t] \times [0, s]$. Again, $s = t > \frac{1}{2}$ so $\sigma([0, t] \times [0, s]) = s + t > 1 = \sigma([0, 1] \times [0, st])$ which contradicts optimality and we are done. □

The above solution of the EIP on $[0, 1]^d$ is easily extended to $[0, n_1] \times [0, n_2] \times \ldots \times [0, n_d]$ where the n_i may be any positive real numbers. However, it does more than just give approximate results for $P_{n_1} \times P_{n_2} \times \ldots \times P_{n_d}$ since there is an embedding of $P_{n_1} \times P_{n_2} \times \ldots \times P_{n_d}$ into $[0, n_1] \times [0, n_2] \times \ldots \times [0, n_d]$: Define ζ pointwise by

$$\zeta(i_1, i_2, \ldots, i_d) = [i_1 - 1, i_1] \times [i_2 - 1, i_2] \times \ldots \times [i_d - 1, i_d]$$

and for any ideal S of $P_{n_1} \times P_{n_2} \times \ldots \times P_{n_d}$

$$\zeta(S) = \bigcup_{i \in S} \zeta(i).$$

$\zeta(S)$ is then an ideal in $[0, n_1] \times [0, n_2] \times \ldots \times [0, n_d]$ such that $\lambda(\zeta(S)) = |S|$ and $\sigma(\zeta(S)) = |\Theta(S)|$. For example if

$$S = \{(1,1), (1,2), (1,3), (1,4), (2,1), (2,2), (3,1)\} \subseteq P_3 \times P_5,$$

then $\zeta(S) \subseteq [0, 3] \times [0, 5]$ is cross-hatched in Fig. 10.2. The elements of S itself are circled. The embedding shows that for all values of k between 0 and $\prod_{i=1}^{d} n_i$ the generalized $F(k)$ gives a lower bound for $\min_{|S=k|} |\Theta(S)|$. The lower bound is actually sharp for many values of k, including all $k = \prod_{i=1}^{d-1} n_i \times k'$, $\frac{n_{d-1}}{4} \le k' \le n_d - \frac{n_{d-1}}{4}$.

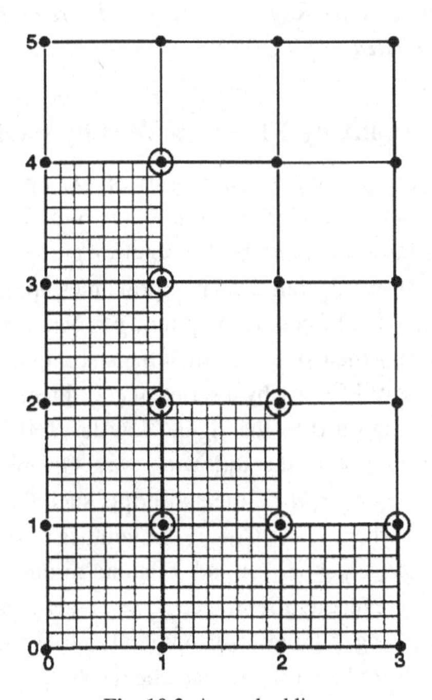

Fig. 10.2 An embedding.

10.2 The Kleitman–West problem

The *Johnson graph*, $J(d, n)$, has $V = \{v \in \{0,1\}^d : \sum v_i = n\}$ and $E = \{\{v, w\} : |\{i : v_i \neq w_i\}| = 2\}$. This name is taken from [**24**], but $J(d, n)$ is also commonly referred to as "the Hamming sphere of dimension d and radius n" or "the nth slice of Q_d". Kleitman [**62**], [**63**] was the first to point out that the EIP on $J(d, n)$ does not have nested solutions. Leader also mentioned it in his survey of combinatorial isoperimetric problems [**69**] as an important unsolved problem. The author first heard about the problem from D. West and called it the Kleitman–West problem. The author's paper [**49**] on the Kleitman–West problem was independent of that of Bollobas and Leader on the EIP for $P_{n_1} \times P_{n_2} \times ... \times P_{n_d}$ (both papers appeared in 1991) but used the same basic strategy of passing to a continuous limit. The Kleitman–West problem presents an additional hurdle in that it not only lacks nested solutions, but there is no obvious way to pass to a continuous limit. However, the theory of stabilization, as laid out in Chapters 3 and 5, shows the way. Note that $J(d, n)$ is the subgraph of $Q_d^{(2)}$, the 2-pather (see Section 5.6 of Chapter 5) of the graph of the d-cube, induced by its vertices. One deduces then that the reflective symmetries, $\mathcal{R}_{i,j}$, of Q_d (which interchange the ith and jth coordinates and therefore preserve $J(d, n)$), are extended stabilizing on $J(d, n)$.

Exercise 10.1 *Show that the $\mathcal{R}_{i,j}$, $1 \leq i < j \leq d$, are actually stabilizing on $J(d, n)$ (not just extended stabilizing).*

10.2.1 Simplifying Kleitman–West by stabilization

The $\mathcal{R}_{i,j}$ are the same reflective symmetries of Q_d upon which Bernstein, Steiglitz and Hopcroft [**14**] based their notion of two-dimensional stability that led to the general definition of stability. The Johnson graph, upon which the $\mathcal{R}_{i,j}$ operate, is different from Q_d, but the same principles apply (see Sections 3.2 and 5.3). If we choose the Fricke–Klein point, $p \in \mathbb{R}^d$, so that its coordinates are decreasing (if $i < j$ then $p_i > p_j$) then the stability order of $J(d, n)$ has $x < y$ iff x can be derived from y by a series of coordinate interchanges which move a 1 to the left and a 0 to the right. Now let f be a mapping on the vertices of $J(d, n)$ defined in the following way. Given $x = (x_1, x_2, ..., x_d)$ let $1 \leq \beta_1 \leq \beta_2 \leq ... \leq \beta_n \leq d$ be the subscripts such that $x_{\beta_j} = 1$ and then $f(x) = (i_1, i_2, ..., i_n)$ where $i_j = \beta_j - j$, the number of 0's to the left of x_{i_j}. Thus $0 \leq i_1 \leq i_2 \leq ... \leq i_n \leq d - n$ and f is one-to-one and onto the set of n-tuples satisfying these inequalities (since $\beta_j = i_j + j$ locates the jth 1 in x).

If these nondecreasing n-tuples with integral entries between 0 and $m = d - n$ are partially ordered coordinatewise, the resulting poset is known in the combinatorial literature as $L(n,m)$ and $f : \mathcal{S}(J(d,n)) \to L(n,m)$ is an order isomorphism. But how does the boundary functional, $|\Theta(S)|$, transform under f? Since $J(d,n)$ is regular and each vertex is incident to $n(d-n)$ edges,

$$|\Theta(S)| = n(d-n) - 2\,|\{\{x,y\} \in E : x, y \in S\}|\,.$$

If S is stable, i.e. an ideal in $\mathcal{S}(J(d,n))$, then the edges of $J(d,n)$ with both ends in S correspond to pairs (x,y) and $i < j$ with $x_i = 1$, $y_i = 0$, $x_j = 0$, $y_j = 1$ and $x_k = y_k$ for all $k \neq i, j$. $x <_{\mathcal{S}} y$ so if $y \in S$ then $x \in S$. The number of such edges for each $y \in S$ is thus counted by $r(y') = \sum_{i=1}^{n} y'_i$ where $y' = f(y)$. Note that $r(y')$ is also the rank of y' in $L(n,m)$. Thus

$$|\Theta(S)| = nm - 2\sum_{x \in S} r(f(x)),$$

and we have the following theorem.

Theorem 10.2 *Stabilization reduces the EIP on $J(d,n)$ to the maximum rank ideal problem on $L(n,m)$.*

Thus, given l, $0 \leq l \leq \binom{m+n}{n}$, we seek to maximize $r(S) = \sum_{y \in S} r(y)$ over all ideals, S, of $L(n,m)$. This reduction of the Kleitman–West problem is not only simpler computationally, but leads to further insights. Note that $L(n,m)$ is self-dual, where $(i_1, i_2, \ldots, i_n)^* = (m - i_n, m - i_{n-1}, \ldots, m - i_1)$. If S is an ideal of $L(n,m)$ then $S^c = L(n,m) - S$ is a filter and $(S^c)^*$ is an ideal again. Also

$$r(S) + r(S^c) = r(L(n,m))$$

and

$$r(S^*) = \sum_{y \in S^*} r(y) = \sum_{y \in S} nm - r(y) = nm\,|S| - r(S)\,.$$

Thus if an ideal S maximizes $r(S)$ for its cardinality, then so does its dual-complement, $(S^c)^*$.

Example 10.1 *The EIP on $J(5,2)$ (see Fig. 10.3) reduces to the maximum weight ideal problem on $L(2,3)$, whose Hasse diagram is shown if Fig. 10.4. Table 10.1 lists all ideals in $L(2,3)$ of size $l = 0, \ldots, 4$ and the corresponding rank.*

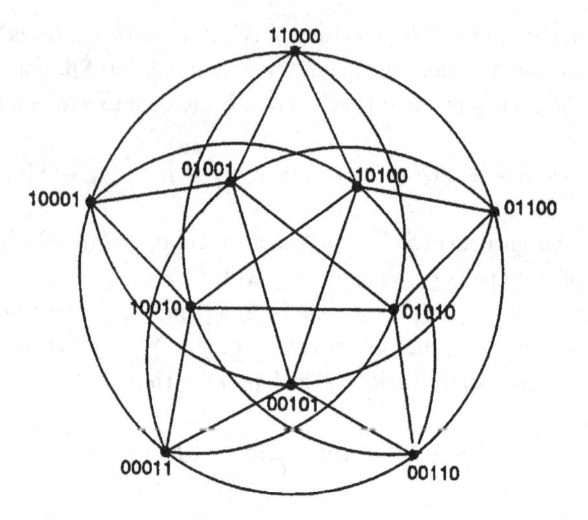

Fig. 10.3 The Johnson graph, $J(5, 2)$.

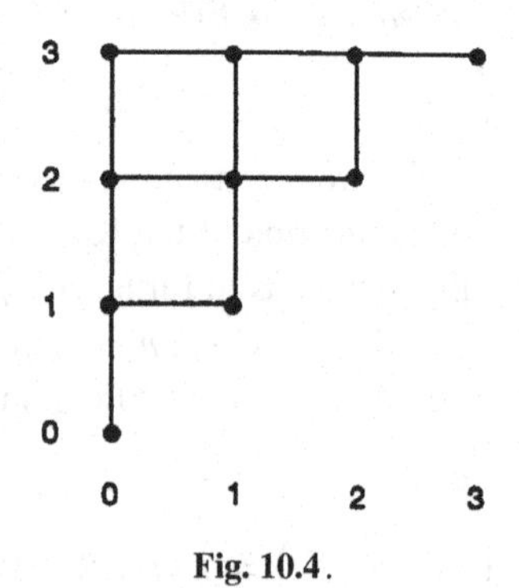

Fig. 10.4.

Table 10.1.

l	S	$r(S)$
0	$\emptyset$	0
1	$\{(0, 0)\}$	0
2	$\{(0, 0), (0, 1)\}$	1
3	$\{(0, 0), (0, 1), (0, 2)\}$	3
	$\{(0, 0), (0, 1), (1, 1)\}$	3
4	$\{(0, 0), (0, 1), (0, 2), (0, 3)\}$	6
	$\{(0, 0), (0, 1), (0, 2), (1, 1)\}$	5

Thus for $l = 4$, $S_4 = \{(0,0), (0,1), (0,2), (0,3)\}$ is the unique ideal in $L(2,3)$ maximizing rank. By the observations about dual-complements above, $S_6 = (S_4^c)^*$ is the unique rank-maximizing ideal for $l = 10 - 4 = 6$. But

$$S_6 = \{(0,0), (0,1), (0,2), (1,1), (1,2), (2,2)\}$$

does not contain S_4 as a subset. Thus there is no way to extend the solution for $l = 4$ to one for $l = 6$, so the maximum rank ideal problem on $L(2,3)$ does not have nested solutions. By the theory of stabilization, the original Kleitman–West problem on $J(5,2)$ also fails to have nested solutions.

Problem 10.1 *Prove or disprove that $L(n,m)$ is Macaulay (see Chapter 8, Section 8.1.5).*

With the reduction to the maximum rank ideal problem on $L(n,m)$, it becomes apparent how to pass to a limit. Since $J(d,n)$ is isomorphic to $J(d, d-n)$, $L(n,m)$ is isomorphic to $L(m,n)$ so we may always assume that $m \geq n$ and let $m \to \infty$ (keeping n fixed). Then

$$\lim_{m\to\infty} \frac{L(n,m)}{m^n} = \left\{x \in [0,1]^n : \forall j, x_j \leq x_{j+1}\right\},$$

which we call $L(n)$. Geometrically $L(n)$ is a simplex and since every $x \in [0,1]^n$ with distinct coordinates can be corresponded to a unique element in the interior of $L(n)$ by a permutation of its coordinates, the volume of $L(n)$ is $1/n!$. The continuous limit of the maximum rank ideal problem on $L(n,m)$ is: Given v, $0 \leq v \leq 1/n!$, to maximize the rank, $r(S) = \int_S r(x)\,dx$, over all closed ideals of $L(n)$ of volume $\lambda(S) = v$.

Conjecture 10.1 *The critical ideals for the maximum rank ideal problem on $L(n)$ are*

$$S_j(t) = \left\{x \in L(n) : x_j \leq t\right\},\ j = 1, ..., n,$$

where t is determined by $\lambda\left(S_j(t)\right) = v$.

First we show how this conjecture leads to a solution of the maximum rank ideal problem on $L(n)$. Recall that for $x \in L(n)$, $\overleftarrow{x} = \{y \in L(n) : y \leq x\}$, the principal ideal generated by x. By direct integration (see [**49**] for details)

$$\int_{\overleftarrow{x}} dy = \frac{1}{n!}\sum_{\sigma}\binom{n}{\sigma}\prod_{j=1}^{n}(\Delta x_j)^{\sigma_j}$$

where Δ is the difference operator $\Delta x_j = x_j - x_{j-1}$ with $x_0 = 0$, the sum is over all $\sigma \in (\mathbb{Z}_+^n)$ such that $\sum_{i=1}^{j}\sigma_i \geq j$ with equality for $j = n$, and $\binom{n}{\sigma}$ is a multinomial coefficient. Also

$$\int_{\overleftarrow{x}} r(y)\,\mathrm{d}y = \frac{1}{2n!}\sum_{\sigma}\left[\sum_{i=1}^{n}\left[\left(2\sum_{k=i+1}^{n}\sigma_k\right)+\sigma_i\right]x_i\right]$$
$$\times\binom{n}{\sigma}\prod_{j=1}^{n}(\Delta x_j)^{\sigma_j}.$$

By substitution, since

$$S_j(t) = \overleftarrow{x} \text{ with } x_i = \begin{cases} t & \text{if } i \leq j \\ 1 & \text{if } i \geq j, \end{cases}$$

$$\int_{S_j(t)} \mathrm{d}y = \frac{1}{n!}\sum_{k\geq j}\binom{n}{k}t^k(1-t)^{n-k}$$

and

$$\int_{S_j(t)} r(y)\,\mathrm{d}y = \frac{1}{n!}\sum_{k\geq j}(nt+n-k)\binom{n}{k}t^k(1-t)^{n-k}.$$

If we let $v' = n!v$ (so $0 \leq v' \leq 1$), for $S_j(t)$, then t is determined by

$$n!\lambda\left(S_j(t)\right) = \sum_{k\geq j}\binom{n}{k}t^k(1-t)^{n-k} = v'$$

and

$$n!r\left(S_j(t)\right) = nv' - \sum_{k\geq j}(k-nt)\binom{n}{k}t^k(1-t)^{n-k}.$$

Thus, given v', maximizing $r\left(S_j(t)\right)$ is equivalent to minimizing

$$R_j\left(v'\right) = \sum_{k\geq j}(k-nt)\binom{n}{k}t^k(1-t)^{n-k}.$$

See Fig. 10.5.

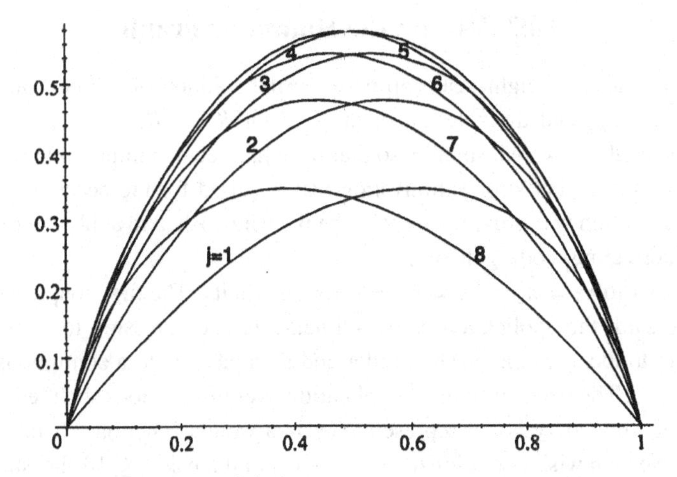

Fig. 10.5 Graphs of $R_j(v')$, the critical Kleitman–West functions, $n = 8$.

Unlike the EIP on $[0, 1]^n$, the critical functions for the EIP on $L(n)$ change significantly with n. If we let $T = F^{-1}(1 - v')$ then

$$R_j(v') = \sum_{k \geq j} (k - nt) \binom{n}{k} t^k (1 - t)^{n-k}$$
$$\simeq \frac{1}{\sqrt{2\pi}} \int_T^\infty \sigma s e^{-\frac{s^2}{2}} ds$$
$$= \frac{\sigma e^{-\frac{T^2}{2}}}{\sqrt{2\pi}}$$

where $\mu = nt, \sigma = \sqrt{nt(1-t)}, k \simeq \mu + \sigma s$ and $j \simeq \mu + \sigma t$. T depends only on v' so for fixed v' the only dependence of our approximate objective function is through $\sigma = \sqrt{nt(1-t)}$. Since t is increasing in j, σ is unimodal in j and its minimum must occur at the extremes, $j = 1$ or n.

In [**49**] the author claimed to prove Conjecture 10.1. The argument is variational, coming down to the claim that if the first variation is 0 then the second variation is > 0 so there can be no interior maximum. However, the "proof" is only the calculation of the second variation in a significant special case and does not cover all possibilities. Several months of effort in the preparation of this monograph failed to fill the logical gap and the proof must be declared in default. However, we still believe the conjecture is true.

10.3 VIP on the Hamming graph

There was another eight-year gap between the solution of the EIP on $P_{n_1} \times P_{n_2} \times ... \times P_{n_d}$ and the solution of the VIP on $K_{n_1} \times K_{n_2} \times ... \times K_{n_d}$. This time the methods were available, so that what had seemed impossible was now merely difficult, but strong motivation was required for the necessary effort. That motivation was provided by R. Urbanke who wished to apply the result in (error-correcting) coding theory.

Restrict to $n_1 = n_2 = ... = n_d = n$ for simplicity. The first couple of steps are standard, the application of one-dimensional compression to restrict the problem to ideals in the product order and then passage to a continuous limit as $n \to \infty$. The one additional complication over the previous such reductions is that the boundary is not representable by a weight function. Given v, $0 \leq v \leq 1$, we then wish to minimize $\lambda(\Phi(S))$ over all ideals $S \subseteq [0, 1]^d$ such that $\lambda(S) = v$. Again, like the EIP on $J(d, n)$, solutions of this problem are not so obvious. For small values of v (near 0) the optimal ideal will be a cube, $[0, t]^d$. For large values of v (near 1), the remarks following Theorem 10.2 shows it will be the dual-complement of $[0, t]^d + \Phi\left([0, t]^d\right)$. For $d = 2$ this is just another subcube (square) and one can easily see that the square is optimal for all v. For $d = 3$, however, the dual-complement of $[0, t]^d + \Phi\left([0, t]^d\right)$ is $[0, 1 - t]^d + \Phi\left([0, 1 - t]^d\right)$, a subcube with arms, one extending in each coordinate direction to the upper boundary of the unit cube. If this monstrous set (at first we called it *the Hydra*) is optimal for v sufficiently large (as it must be), then what other monsters will be optimal in dimensions higher than 3, for intermediate values of v? However, as in the myths of old, understanding the nature of a monster can render it harmless, even convert it into a useful friend.

A simple variational argument shows that an optimal ideal $S \subseteq [0, 1]^d$ must have the following form: For $1 \leq i \leq d$ let

$$t_i = \max\left\{x_i : x \in S \text{ but } \exists \epsilon > 0 \text{ such that } x + \epsilon\delta^{(i)} \notin S\right\},$$

where

$$\delta_j^{(i)} = \begin{cases} 1 & \text{if } j = i \\ 0 & \text{if } j \neq i. \end{cases}$$

This determines an order-preserving map $f : [0, 1]^d \to \{0 < 1\}^d$ by $f((x_1, x_2, ..., x_n)) = (y_1, y_2, ..., y_n)$ where

$$y_i = \begin{cases} 0 & \text{if } x_i \leq t_i \\ 1 & \text{if } x_i > t_i. \end{cases}$$

Then $f(S) = S'$ is an ideal of $\{0 < 1\}^d$ and $S = f^{-1}(S')$.

From there one can guess that the Hydra and its spawn are just the ideals corresponding to Hamming balls (centered at 0^d),

$$HB(r,d) = \left\{ y \in \{0,1\}^d : \sum y_i \le r \right\},$$

with $t_1 = t_2 = \ldots = t_d = t$. Since $f^{-1}(HB(r,d))$ with parameter t is a set in the continuous Hamming graph, but not itself a Hamming ball, we call it the *quotient Hamming ball, $QHB(r,d,t)$*.

Theorem 10.3 *The critical ideals for the VIP on $[0,1]^d$ are the quotient Hamming balls, $QHB(r,d,t)$, for $r = 0,\ldots,n-2$, where t is determined by*

$$\lambda(QHB(r,d,t)) = v.$$

A proof is given in [**50**]. Its logic is qualitatively different from that for the EIP on $[0,1]^d$ which is based on convexity. The heart of the proof is a new Steiner operation based on reflective symmetry.

It is shown in [**50**] that

$$\lambda(QHB(r,d,t)) = \sum_{i=0}^{r} \binom{d}{i} t^{d-i}(1-t)^i$$

and

$$\lambda(\Phi(QHB(r,d,t))) = \binom{d}{r+1} t^{d-r-1}(1-t)^{r+1}.$$

Together these equations parametrically define the function $\lambda(\Phi(QHB(r,d,v)))$ (see Fig. 10.6). Every $QHB(r,d,v)$ is optimal for some interval of v's, but only those with r near $(d-2)/2$ are significant. As with the Bollobas–Leader result, the embedding of $(K_n)^d$ into $[0,n]^d$ shows that for all k, $0 \le k \le n^d$,

$$\min_{\substack{S \subseteq (K_n)^d \\ |S|=k}} |\Phi(S)| \ge n^d \times \min_r \lambda\left(\Phi\left(QHB\left(r,d,\frac{k}{n^d}\right)\right)\right)$$

and the inequality is sharp for many values of k.

10.4 Sapozhenko's problem

A. A. Sapozhenko expressed an interest in the solution of the VIP on $J(d,n)$ so that he might apply it in his work on Dedekind's problem (estimating the number of antichains in the Boolean lattice, $\{0<1\}^n$). I. Leader also mentions

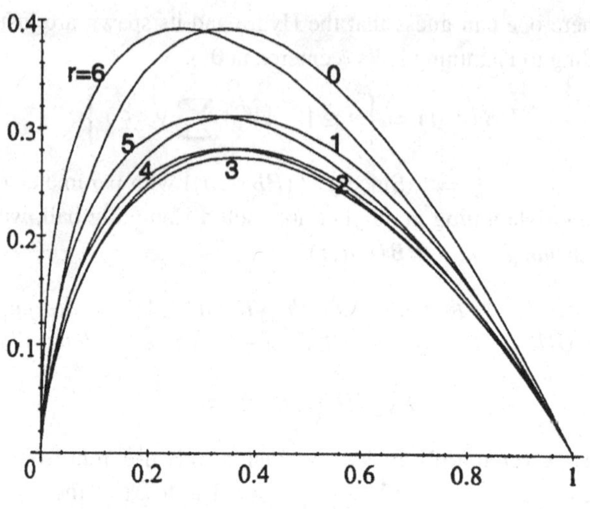

Fig. 10.6 Graphs of $\lambda\left(\Phi\left(QHB\left(r, 8, v\right)\right)\right)$.

it in [**69**]. The solutions are again not nested, and it is not even clear what they are. Stabilization (see Section 10.2) again reduces the problem to ideals in $L(n, m)$ (as it did with the Kleitman–West problem, but there is no weight function representing $|\Phi(S)|$, and the neighborhood of a vertex in $L(n, m)$ is a bit strange.

10.4.1 The crooked neighborhood of $v \in L(n, m)$

The neighbors of v are not just those $w \in L(n, m)$ which differ from it in exactly one coordinate. The relationship between the corresponding vertices, v' and w' of $J(d, n)$, is that they differ in exactly two coordinates, say the ith and jth with $i < j$, and then (assuming $v < w$) $v'_i = 1$, $w'_i = 0$ but $v'_j = 0$, $w'_j = 1$. If there are no 1s in the coordinates of v' and w' between the ith and jth, then the corresponding v and w only differ in one coordinate. However, if there are 1s in those intervening coordinates, then the relationship is a bit more complicated. The isomorphism of the stability order of $J(d, n)$ to $L(n, m)$ takes the d-tuple, v', of 0s and 1s to the nondecreasing n-tuple, v, of natural numbers ($n = d - m$) whose kth entry is the number of 0s to the left of the kth 1 (counting from the left) in v'. Thus we can go up from v to neighbors of v by raising the kth entry, v_k, until it equals v_{k+1}. Then we can go on to other neighbors of v by leaving v_k at the value of v_{k+1} and raising v_{k+1} (assuming $v_{k+1} < v_{k+2}$ or $k + 1 = n$ and $v_{k+1} < m$) and so on. In general $v < w$ are neighbors in $L(n, m)$ if w can be obtained from v by removing its kth coordinate, v_k, and inserting a new lth coordinate, w_l, $l \geq k$.

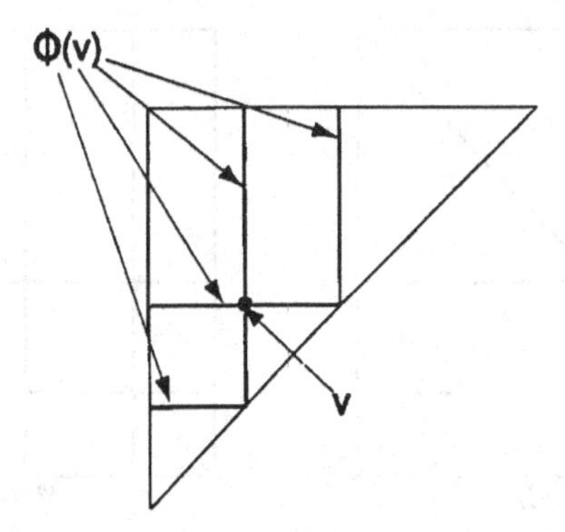

Fig. 10.7.

Passing to the continuous limit ($m \to \infty$, n fixed) gives the corresponding problem on $L(n)$. The crooked neighborhood of a point in $L(2)$ is illustrated in Fig. 10.7.

One can then guess that the critical ideals are the same as they were for the continuous limit of the VIP on $J(d, n)$,

$$S_j(t) = \left\{x \in L(n) : x_j \leq t\right\}, j = 1, ..., n.$$

Actually, the first one is superfluous since $S_1(t) + \Phi(S_1(t)) = L(n)$. From past experience, one would expect proving this conjecture to be difficult, combining the technical challenges presented by the EIP on $L(n)$ and the VIP on $[0, 1]^n$. Instead, we receive a gift from the gods.

10.4.2 Dido's principle

It is well known (see [**60**]) that the curve of fixed length, l, together with a straight line segment of indeterminate length, that bound the maximum area, A, is a semicircle and its diameter. This may be deduced from the classical isoperimetric theorem by reflecting the curve about the straight line to complete a simple closed curve of length $2l$ enclosing area $2A$. The origin of this idea has been attributed to Queen Dido in a myth about the founding of Carthage (see Appendix A.2).

10.4.2.1 The Didonean embedding of $L(n)$ into $[0, 1]^n$

Given $S \subseteq L(n)$, define

$$\Delta(S) = \bigcup_{\pi \in \mathfrak{S}_n} \pi(S),$$

$\mathfrak{S}_n$ being the symmetric group acting on the coordinates of $L(n)$.

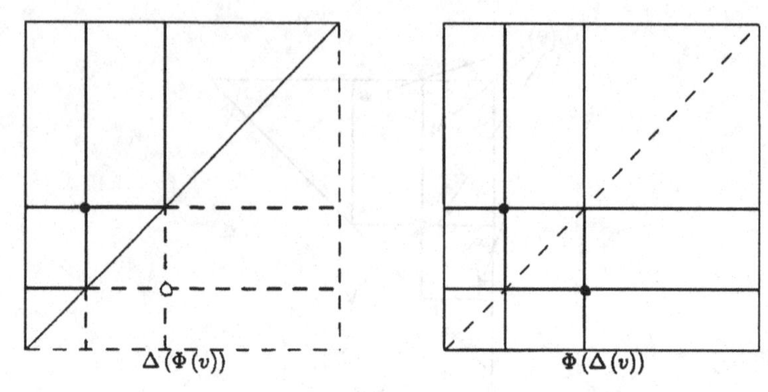

Fig. 10.8.

Lemma 10.1 *If S is an ideal of $L(n)$ then $\Delta(S)$ is an ideal of $[0,1]^n$.*

Proof If $x' < y'$ and $y' \in \Delta(S)$, let x and y be the nondecreasing rearrangements of x' and y' respectively. So there exist $\pi, \sigma \in \mathfrak{S}_n$ such that $x' = \pi(x)$ and $y' = \sigma(y)$. Since $y' \in \Delta(S)$, $y \in S$. x' only differs from y' in one coordinate, where it is less, so $x < y$ in $L(n)$. Thus $x \in S$ and so $x' = \pi(x) \in \Delta(S)$. □

Lemma 10.2 *For any ideal S' of $[0,1]^n$, $\forall \pi \in \mathfrak{S}_n$, $\pi(S') = S'$, if and only if $S' = \Delta(S' \cap L(n))$.*

Lemma 10.3 $\Delta(\Phi(S)) = \Phi(\Delta(S))$.

Proof Fig. 10.8 illustrates how Δ straightens the crooked neighborhood of $v \in L(2)$. □

Lemma 10.4 *For any ideal $S \subseteq L(n)$,*

(1) $\lambda(\Delta(S)) = n!\lambda(S)$,
(2) $\lambda(\Phi(\Delta(S))) = n!\lambda(\Phi(S))$ *and*
(3) $\Delta(S_j(t)) = QHB(n, n-j, t)$.

Proof (1) and (2) follow from the fact that the images $\pi(L(n))$ only overlap on their boundaries which are of Lebesgue measure zero. (3) follows from Lemma 10.2 and the observation that $QHB(n, n-j, t)$ is invariant under all $\pi \in \mathfrak{S}_n$ and $QHB(n, n-j, t) \cap L(n) = S_j(t)$. □

Theorem 10.4 *$S_j(t)$ is a solution of the VIP on $L(n)$ if and only if*

$$QHB(n, n-j, t)$$

is a solution of the VIP on $[0,1]^n$.

Proof This follows from Lemma 10.4 and Dido's principle (see Appendix A.2). □

Corollary 10.1 *For each v,* $0 \leq v \leq \frac{1}{n!}$*, some* $S_j(t)$*, with* $\lambda(S_j(t)) = v$*, is a solution of the VIP on* $L(n)$*. Every* $S_j(t)$ *is optimal for some interval of v's, but only those with j near* $(n+2)/2$ *are significant.*

10.5 Comments

10.5.1 Applications

Below are a variety of applications for the problems treated in this chapter.

10.5.1.1 Modeling the brain

Mitchison and Durbin [**78**] came across the wirelength problem for the pairwise product of paths, $P_n \times P_n$, in their investigation of the brains's mapping of the visual field, which is three-dimensional, onto the visual cortex, which is two-dimensional. They hypothesized that the mapping should be optimal in some sense and modeled it as a function, $\eta : P_n \times P_n \times P_n \to P_m \times P_m$, $m = n^{3/2}$, minimizing

$$\sum_{\substack{e \in P_n \times P_n \times P_n \\ \partial(e)=\{x,y\}}} \|\eta(x) - \eta(y)\| .$$

Unable to solve that problem, they simplified it to $\eta : P_n \times P_n \to P_m$, $m = n^2$, minimizing

$$\sum_{\substack{e \in P_n \times P_n \\ \partial(e)=\{x,y\}}} \|\eta(x) - \eta(y)\| ,$$

which is the wirelength problem. Their solution optimally interpolates between the Bollobas–Leader solutions of the EIP for $P_n \times P_n$. Fishburn, Tetali and Winkler [**36**] recently extended the Mitchison–Durbin result to $P_{n_1} \times P_{n_2}$ but evidently it had already been achieved by Muradian and Piloposian [**80**] in 1980 (in Armenian). All extensions to higher dimensions, including the original problem of Mitchison and Durbin, remain open. It is fairly easy to formulate a conjecture for an optimal numbering (when the range is a path, P_m) but a proof will be hard.

10.5.1.2 Multicasting to minimize noise

Berger-Wolf and Reingold [**12**] proposed the bandwidth of the Hamming graph, $BW\left(K_{n_1} \times K_{n_2} \times \ldots \times K_{n_d}\right)$, as a measure of the effect of noise in the multicasting of ordered data. Theorem 4.6 (in Chapter 4) and Theorem 10.3 of Section 10.3 combine to show that (assuming d and n are even)

$$\begin{aligned} BW\left(K_n \times K_n \times \ldots \times K_n\right) &\geq \binom{d}{d/2} \frac{n^d}{2^d} \\ &\simeq \sqrt{\frac{2}{\pi d}} n^d \text{ as } d \to \infty \end{aligned}$$

by Stirling's formula. In [**51**] a numbering, $\overline{\eta}$, is given that achieves this lower bound for $d = 2$ and asymptotically as $d \to \infty$.

10.5.1.3 Dedekind's problem again

We have already mentioned Sapozhenko's request for a solution of the VIP on the Johnson graph, $J(d, n)$ for application to Dedekind's problem. Unfortunately, the asymptotic analysis we presented in Section 10.4 does not apply to the cases that are of greatest interest for his application. In order for the Didonean embedding to work we must have $m = d - n \to \infty$, which means $d \to \infty$, n being held fixed. The analysis can be extended to allow $n \to \infty$ slowly (i.e. $\frac{n}{d} \to 0$) and by duality if $\frac{n}{d} \to 1$. However, technical problems arise if $\frac{n}{d}$ is on the order of $\frac{1}{2}$.

10.5.1.4 The doctor's waiting room problem

There are a number of unsolved combinatorial problems to which the analytic methods of the chapter may apply, but the technical challenges are formidable. One of the most intriguing arises in queueing theory: d doctors share a waiting room having a total capacity R. Patients arrive and depart at random (Poisson processes with different parameters for each doctor). Sometimes patients must be turned away (e.g. if the waiting room is full) and sometimes a doctor may be idled by not having any of his patients waiting when he is available. We wish to find a policy for admitting patients to the waiting room which will minimize the average idle time for doctors. Foschine and Gopinath [**37**] have formulated this problem as follows. Let

$$K(R) = \left\{a \in \mathbb{Z}_+^d : \sum a_i \leq R\right\}$$

be partially ordered coordinatewise. Given $c_1, c_2, \ldots, c_d > 0$ and $w_1,$

$w_2, \ldots, w_d > 0$, for $a \in K(R)$ let

$$W(a) = \sum c_i w_i^{a_i}$$

and for any $A \subseteq K(R)$ let

$$W(A) = \sum_{a \in A} W(a).$$

Also let $A_i = \{a \in A : a_i = 0\}$. Then the DWRP is to minimize the ratio

$$\frac{\sum_i W(A_i)}{W(A)}$$

over all ideals, $A \subseteq K(R)$. Foschini and Gopinath have shown that for $d \leq 3$ there exists a function $k : \{0,1\}^d \to \mathbb{Z}_+$ such that the ideal

$$A = \left\{a \in K(R) : \sum a_i x_i \leq k(x) \, \forall x \in \{0,1\}^d\right\}$$

is optimal and conjecture that such solutions exist for all d.

10.5.2 The Hwang–Lagarias theorem

There is one other solution (besides the aforementioned result of Mitchison–Durbin, Fishburn–Tetali–Winkler and Muradian–Piloposian) of a wirelength-type problem for a graph (actually a hypergraph) which does not have nested solutions for the EIP. Let V be the vertex-set and $E \subseteq 2^V$ be the edge-set of a hypergraph, H, i.e. E is any set of subsets of V. Given any numbering, η, of V, its *wirelength* is

$$wl(\eta) = \sum_{e \in E} [\max_{v \in e} \eta(v) - \min_{v \in e} \eta(v)]$$

and then

$$wl(H) = \min_{\eta} wl(\eta).$$

For $S \subseteq V$, let

$$\Theta(S) = \{e \in E : \exists v, w \in e, v \in S \& w \notin S\}.$$

As before,

$$wl(\eta) = \sum_{\ell=0}^{|V|} |\Theta(S_\ell(\eta))|.$$

F. K. Hwang and J. C. Lagarias [**57**] solved the EIP for the hypergraph whose vertices are the k-sets of an n-set (the same as the Johnson graph, $J(n, k)$) and whose edges are given by the elements of the n-set, $\{x_1, x_2, \ldots, x_n\}$, i.e.

$$E = \{\{v \in V : x_i \in v\} : 1 \leq i \leq n\},$$

showing that it does not have nested solutions. They then solved the wirelength problem for this hypergraph. Their strategy is similar to that of Mitchison *et al.* but with a clever use of duality and inclusion–exclusion.

10.5.3 Discontinuous Steiner operations

Because the problems treated in this chapter do not have nested solutions, the operations involved in their ultimate solutions cannot have the third property of a Steiner operation: $\forall S \subseteq T,\ SteinOp(S) \subseteq SteinOp(T)$, which we have called *continuity*. The MWI-morphisms of Chapter 9 are also discontinuous in this sense.

10.5.4 Kleitman–West again

The exquisite delicacy of the solutions to the Kleitman–West problem is illustrated by the counterexample that Ahlswede and Cai [**4**] found to a natural conjecture of Kleitman.

Afterword

Almost forty years ago I was persuaded that combinatorial isoperimetric problems were worthy of systematic investigation. The edge- and vertex-isoperimetric problems were clearly fundamental aspects of graph theory. They had already been applied to the wirelength and bandwidth problems on d-cubes and other graphs which had engineering implications. As analogs of the classical isoperimetric problem of Greek geometry they seemed certain to lead to further useful results. Over the years this analogy, with the pressure of prospective applications, has produced profound solution methods; spectral, global and variational.

It has been very difficult to bring closure to the writing of this monograph since every time I go over the material, new insights appear and demand to be included. Also, tempting new problems keep arising in science, engineering and mathematics itself. For instance, Lubotzky's monograph [**75**] has a whole chapter of unsolved problems. It seems certain that the subject will continue to progress for the foreseeable future, but life is short and we cannot wait until every significant question has been answered. Last week, in a conversation with T. H. Payne, colleague, collaborator and for many years a most reliable source of information about trends in computer science, I mentioned recent work on the profile scheduling problem (see Chapter 8). "Oh, yes," he said with enthusiasm, "that has been applied to optimizing straight-line programs! A 'live variable' must be stored in a register, so the profile equals total storage time. But the latest thing is to minimize *register width*, the maximum number of registers required by a program." Later I realized that register width is equal to (in the notation of Chapter 8)

$$\min_{\eta} \max_{0 \leq k \leq |P|} \left| \Phi^* \left(S_k \left(\eta \right) \right) \right|$$

where the poset **P** represents precedence constraints among the variables and η ranges over all numberings consistent with **P**. I rest my case.

Afterword

Appendix

The classical isoperimetric problem

Let C be a simple closed curve in the plane, of length l enclosing a region, S, of area A. The isoperimetric problem of Greek geometry was to find the maximum value of A for a fixed l, the obvious candidate for a solution being a circle of radius $r = l/2\pi$ and area $A = \pi r^2 = l^2/4\pi$. Since under dilation of the plane by a factor $\kappa > 0$, every such planar set is mapped to a set with boundary of length αl and area $\alpha^2 A$, as a function of l the maximum value of A will be $A(l) = \kappa l^2$ for some constant $\kappa \geq 1/4\pi$. Thus $A(l)$ must be monotone increasing. Minimizing l for fixed A is an equivalent form of the isoperimetric problem whose solution, $l(A) = \sqrt{\kappa A}$, is also monotone increasing. The region S may be assumed convex since, if not, its convex closure would only decrease l and increase A.

A.5 Steiner symmetrization

Let L be any line in the plane and we assume for convenience that it passes through the centroid of S. Take it to be the x-axis for a coordinatization of the plane and let the y-axis also pass through the centroid of S (which is then the origin). Since S is convex, there are functions f_+ and f_- on an interval $[a, b]$ such that

$$S = \{(x, y) : a \leq x \leq b \, \text{and} \, f_-(x) \leq y \leq f_+(x)\} .$$

We may identify S with the pair of functions, $f_\pm$. Then

$$\mathrm{A} = \int_a^b (f_+ - f_-) \, \mathrm{d}x$$

and

$$l = \int_a^b \sqrt{1 + (f_+')^2} \mathrm{d}x + \int_a^b \sqrt{1 + (f_-')^2} \mathrm{d}x.$$

The *symmetrization of S* is given by the pair of functions

$$Symm\,(f_\pm) = \pm \frac{f_+ - f_-}{2}.$$

Therefore

$$\begin{aligned} \mathrm{A}\,(Symm\,(f_\pm)) &= \int_a^b \left[\frac{f_+ - f_-}{2} - \left(- \left(\frac{f_+ - f_-}{2} \right) \right) \right] \mathrm{d}x \\ &= \int_a^b (f_+ - f_-)\,\mathrm{d}x \\ &= \mathrm{A}\,(f_\pm), \end{aligned}$$

and

$$\begin{aligned} l\,(Symm\,(f_\pm)) &= 2 \int_a^b \sqrt{1 + \left(\frac{f_+' - f_-'}{2} \right)^2} \mathrm{d}x \\ &\geq l\,(f_\pm), \text{ by Schwartz's inequality} \end{aligned}$$

with equality iff $f_+ = -f_-$, i.e. S is symmetric wrt L. It is easy to show that S is symmetric wrt every line L iff C is a circle. J. Steiner, the great nineteenth-century mathematician who discovered symmetrization, claimed that this proved the isoperimetric theorem. However his contemporary, Weierstrass, pointed out a logical gap. Steiner's "proof" assumed that there is a solution. Weierstrass gave examples of closely related optimization problems which do not have solutions and supplied a proof that the classical isoperimetric problem does have one, thus closing the gap.

A.6 The legend of the founding of Carthage

The following story is adapted from Virgil's *Aeneid*, Book I, lines 307–72 (pp. 22–23) [**91**]. Long ago Queen Dido and her people sailed out upon the Mediterranean Sea, seeking a new home. They landed on the southern shore (in what is now Libya) and went to the local king, asking for some land: "Just a small piece, what an oxhide will cover." The king agreed and even provided her with a large oxhide. Dido decided they could interpret the king's grant of land more generously than he had intended. They cut the oxhide into a long string and, the next day, used it to "cover" (bound) the landward side of their territory

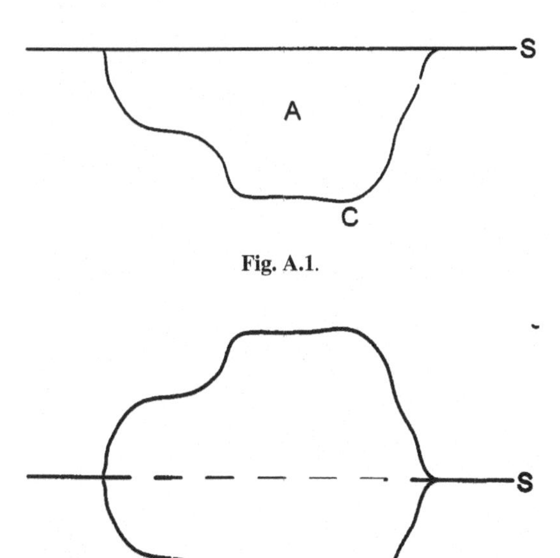

Fig. A.1.

Fig. A.2.

with the shore forming the other side. The city Dido founded, called "Oxhide," was Carthage.

Q: Assuming that the shoreline, *S*, is straight and the oxhide string of a fixed length, *l*, what curve, *C*, should the string follow to "cover" the greatest area (see Fig. A.1)?

A: The optimal curve is a half-circle of radius $r = \frac{l}{\pi}$ with *S* as its diameter. Given any such curve, *C*, of length *l* and bounding area *A*, reflecting *C* about *S* gives a simple, closed curve of length $2l$, surrounding area 2A (see Fig. A.2). By the classical isoperimetric theorem in the plane, any simple closed curve of a fixed length surrounding the maximum area is a circle. Since the image of the given half-circle under the reflection completes a circle, it is optimal.

A.6.1 Dido's principle

If the sets of one isoperimetric problem can be embedded into those of another, essentially preserving area and boundary, and if an optimal set for the second problem is the image of some set for the first, then its preimage is optimal (for the first problem).

References

[1] R. Ahlswede and S. L. Bezrukov, Edge-isoperimetric theorems for integer point arrays, *Appl. Math. Lett.* **8** (1995), 75–80.

[2] R. Ahlswede and N. Cai, On edge-isoperimetric theorems for uniform hypergraphs, preprint 93-018, Diskrete Strukturen in der Mathematik SFB 343 (1993), Universität Bielefeld.

[3] R. Ahlswede and N. Cai, General edge-isoperimetric inequalities, part II: A local–global principle for lexicographic solution, *Europ. J. Combin.* **18** (1997), 479–489.

[4] R. Ahlswede and N. Cai, A counterexample to Kleitman's conjecture concerning an edge-isoperimetric problem, *Combinatorics, Probability and Computing* **8** (1999), 301–305.

[5] R. Ahlswede and G. Katona, Graphs with maximal number of adjacent pairs of edges, *Acta Math. Hungar.* **32** (1978), 97–120.

[6] R. K. Ahuja, T. L. Magnanti and J. B. Orlin, *Network Flows: Theory, Algorithms and Applications,* Prentice-Hall, Englewood Cliffs, NJ (1993).

[7] I. Anderson, *Combinatorics of Finite Sets*, Oxford University Press (1987).

[8] M. C. Azizoğlu and Ö. Eğecioğlu, Extremal sets minimizing dimension-normalized boundary in Hamming graphs, preprint, Department of Computer Science, University of California at Santa Barbara.

[9] R. Bellman, *Dynamic Programming*, Princeton University Press (1957).

[10] C. T. Benson and L. C. Grove, *Finite Reflection Groups*, Bogden and Quigley Inc., Terrytown-on-Hudson, NY (1971).

[11] X. Berenguer and L. H. Harper, Estabilizacion y resulucion de algunos problemas combinatorias en grafos simetricos, *Questiio* **3**(2) (June 1979), 105–117.

[12] T. Y. Berger-Wolf and E. M. Reingold, Index assignment for multichannel communication under failure, *IEEE Trans. Inf. Theory,* **48** (2000), 2656–2668.

[13] A. J. Bernstein, Maximally connected arrays on the n-cube, *SIAM J. Appl. Math.* **15** (1967), 1485–1489.

[14] A. J. Bernstein, K. Steiglitz and J. Hopcroft, Encoding of analog signals for a binary symmetric channel, *IEEE Transactions on Information Theory* **IT-12** (1966), 425–430.

[15] S. L. Bezrukov, On k-partitioning the n-cube, *Proc. Intl. Conf. on Graph Theor.*

Concepts in Comp. Sci., Como, Italy, Springer Verlag Lect. Notes in Comp. Sci. **1197** (1997), 44–55.
[16] S. L. Bezrukov, On an equivalence in discrete extremal problems, *Discr. Math.* **203** (1999), 9–22.
[17] S. L. Bezrukov and A. Blokhuis, A Kruskal–Katona type theorem for the linear lattice, *European J. Combin.* **20** (1999), 123–130.
[18] S. L. Bezrukov, R. Elsässer and S. Das, An edge-isoperimetric problem for powers of the Petersen graph, *Annals of Combinatorics* **4** (2000), 153–169.
[19] S. L. Bezrukov and R. Elsässer, The spider poset is Macaulay, *J. Comb. Theory* **A-90** (2000), 1–26.
[20] S. L. Bezrukov and R. Elsässer, Edge-isoperimetric problems for Cartesian powers of regular graphs, preprint.
[21] G. Birkhoff, *Lattice Theory*, AMS Colloquium Publications, Vol. XXV, Providence, RI, third ed. (1967).
[22] B. Bollobas and I. Leader, Edge-isoperimetric inequalities in the grid, *Combinatorica* **11** (1991), 299–314.
[23] N. Bourbaki, *Groupes et algèbres de Lie*, Hermann, Paris (1968).
[24] A. E. Brouwer, A. M. Cohen and A. Neumaier, *Distance-Regular Graphs*, Springer-Verlag (1989).
[25] T. Carlson, The edge-isoperimetric problem for discrete tori, *Discrete Math.*, **254** (2002), 33–49.
[26] P. Z. Chinn, J. Chvátalová, A. K. Dewdney and N. E. Gibbs, The bandwidth problem for graphs and matrices – a survey, *J. Graph Theory* **6** (1982), 223–254.
[27] F. R. K. Chung, *Spectral Graph Theory*, Reg. Conf. Ser. Math. **92** (1997), AMS.
[28] H. S. M. Coxeter, *Regular Polytopes*, Macmillan, New York (1963).
[29] G. F. Clements and B. Lindström, A generalization of a combinatorial theorem of Macaulay, *J. Combin. Th.* **7** (1969), 230–238.
[30] T. R. Crimmins, H. M. Horwitz, C. J. Palermo and R. V. Palermo, Minimization of mean-square error for data transmitted via group codes, *IEEE Trans. Inf. Th.* **IT-15** (1969), 72–78.
[31] J. Chvátalová, Optimal labelling of a product of two paths, *Discrete Math.* **11** (1975), 249–253.
[32] R. Cypher, Theoretical aspects of VLSI pin limitations, *SIAM J. Computing* **22** (1993), 356–378.
[33] P. Diaconis, *Group Representations in Probability and Statistics*, Inst. of Math. Stat. Lecture Notes **11** (1988).
[34] K. Engel, *Sperner Theory*, Encyclopedia of Mathematics and its Applications **65**, Cambridge University Press (1997).
[35] K. Engel and H-D. O. H. Gronau, *Sperner Theory in Partially Ordered Sets*, Teubner (1985).
[36] P. Fishburn, P. Tetali and P. Winkler, Optimal linear arrangement of a rectangular grid, *Discrete Math.* **213** (2000), 123–139.
[37] G. J. Foschini and B. Gopinath, Sharing memory optimally, preprint, Bell Labs, Murray Hill, New Jersey.
[38] C. Fox, *An Introduction to the Calculus of Variations*, Dover Publications (1987).
[39] M. R. Garey and D. S. Johnson, *Computers and Intractability: A Guide to the Theory of NP-Completeness*, W. H. Freeman (1979).

[40] H. H. Goldstine, *A History of the Calculus of Variations from the 17th through the 19th Century*, Springer Verlag (1980).
[41] S. W. Golomb, *Shift Register Sequences* (rev. ed.), Aegean Park Press (1982).
[42] C. Greene, Posets of shuffles, *J. Combin. Theory Ser.* A **47** (1988), 191–206.
[43] F. Harary, The maximum connectivity of a graph, *Proc. Nat. Acad. Sci. USA* **46** (1962), 1142–1146.
[44] F. Harary and H. Harborth, Extremal animals, *J. Combin. Inform. and System Sci.* **1** (1976), 1–8.
[45] L. H. Harper, Optimal assignments of numbers to vertices, *J. SIAM* **12** (1964), 131–135.
[46] L. H. Harper, Optimal numberings and isoperimetric problems on graphs, *J. Comb. Th.* **1** (1966), 385–393.
[47] L. H. Harper, Chassis layout and isoperimetric problems, *Jet Propulsion Lab. Space Proj. Summary* **11** (1970), 37–66.
[48] L. H. Harper, Stabilization and the edgesum problem, *Ars Combinatoria* **4** (1977), 225–270.
[49] L. H. Harper, On a problem of Kleitman and West, *Disc. Math.* **93** (1991), 169–182.
[50] L. H. Harper, On an isoperimetric problem for Hamming graphs, *Disc. Appl. Math.* **95** (1999), 285–309.
[51] L. H. Harper, On the bandwidth of a Hamming graph, *Theor. Computer Sci.* **301** (2003), 491–498.
[52] L. H. Harper and D. Dreier, The edge-isoperimetric problem on the 600-vertex regular solid, *Linear Algebra Appl.* **368** (2003), 209–228.
[53] L. H. Harper and F. Hergert, The isoperimetric problem in finite projective planes, *Congressus Numerantium* **103** (1994), 225–232.
[54] H. Hiller, *Geometry of Coxeter Groups*, Pitman, Marshfield, MA (1982).
[55] D. A. Holton and J. Sheehan, *The Petersen Graph*, Australian Math. Soc. Lecture Series **7,** Cambridge University Press (1993).
[56] J. E. Humphries, *Reflection Groups and Coxeter Groups*, Cambridge Studies in Adv. Math. **29,** Cambridge University Press (1990).
[57] F. K. Hwang and J. C. Lagarias, Minimum range sequences of all k-subsets of a set, *Discrete Math.* **19** (1977), 257–264.
[58] G. Katona, A theorem of finite sets, in *Theory of Graphs*, proceedings of a conference held at Tihany in 1966, Akademiai Kiado (1968), pp. 187–207.
[59] W. H. Kautz, Optimized data encoding for digital computers, *Convention Record I. R. E.* (1954), pp. 47–57.
[60] N. D. Kazarinoff, *Geometric Inequalities*, Random House (1961).
[61] R. Klasing, B. Monien, R. Peine and E. A. Stöhr, Broadcasting in butterfly and de Bruijn networks. *Proceedings International Workshop on Broadcasting and Gossiping 1990* (Sechelt, BC), *Discrete Appl. Math.* **53** (1994), 183–197.
[62] D. J. Kleitman, Extremal hypergraph problems, in *Surveys in Combinatorics, Proceedings of the 7th British Combinatorial Conference* (ed. B. Bollobás), vol. 38 of London Math. Society Lecture Notes, Cambridge University Press (1979), pp. 44–65.
[63] D. J. Kleitman, Extremal problems on hypergraphs, in *Extremal Problems for Finite Sets, Visegrad, Hungary* (ed. P. Frankl, Z. Furedi, G. Katona, and D. Miklós), vol. 3 of Bolyai Society Math. Studies (1991), pp. 355–374.

[64] D. J. Kleitman, M. M. Krieger and B. L. Rothschild, Configurations maximizing the number of pairs of Hamming-adjacent lattice points, *Studies in Appl. Math.* **50** (1971), 115–119.

[65] J. B. Kruskal, The number of simplices in a complex, in *Mathematical Optimization Techniques*, University of California Press (1963), pp. 251–278.

[66] J. B. Kruskal, The number of s-dimensional faces in a cubical complex; An analogy between the simplex and the cube, *J. Comb. Th.* **6** (1969), 86–89.

[67] Y.-L. Lai and K. Williams, A survey of solved problems and applications on bandwidth, edgesum and profile of graphs, *J. Graph Theory* **31** (1999), 75–94.

[68] E. L. Lawler, *Combinatorial Optimization: Networks and Matroids*, Holt, Rinehart & Winston (1976).

[69] I. Leader, Discrete isoperimetric inequalities, in *Probabalistic Combinatorics and its Applications* (ed. B. Bollobas *et al.*), *Proc. Symposia Appl. Math.* **44** (1992), AMS, pp. 57–80.

[70] U. Leck, Another generalization of Lindström's theorem on subcubes of a cube, JCT-A **99** (2002), 281–296.

[71] K. Leeb, Salami-Taktik beim Quader-Packen, *Arbeitsberichte des Instituts für Mathematische Maschinen und Datenverarbeitung, Universität Erlangen*, **11** (1978), no. 5, 1–15.

[72] A. Lehman, A result on rearrangements, *Israel J. Math* **1** (1963), 22–28.

[73] J. E. Lindsey III, Assignments of numbers to vertices, *Amer. Math Monthly* **71** (1964), 508–516.

[74] B. Lindström, The optimal number of faces in cubical complexes, *Ark. Mat.* **8** (1971), 245–257.

[75] A. Lubotzky, *Discrete Groups, Expanding Graphs and Invariant Measures*, Birkhäuser (1994).

[76] S. L. MacLane, *Categories for the Working Mathematician*, Springer-Verlag, New York (1971).

[77] F. S. Macaulay, Some properties of enumeration in the theory of modular systems, *Proc. London Math. Soc.* **26** (1927), 531–555.

[78] G. Mitchison and R. Durbin, Optimal numberings of an N×N array, *SIAM J. Algebraic Discrete Methods* **7** (1986), 571–582.

[79] H. S. Moghadam, Compression operators and a solution of the bandwidth problem of the product of n paths, Ph.D. thesis, University of California, Riverside (1983).

[80] D. O. Muradian and T. E. Piliposian, Minimal numbering of the vertices of a rectangular grid (in Russian); *Doklady (Mathematics), Academy of Sciences of the Armenian Soviet Socialist Republic* **70** (1980), 21–27.

[81] R. G. Nigmatullin, *The Complexity of Boolean Functions* (in Russian), Nauka, Moscow, main editorial board for physical and mathematical literature (1990).

[82] V. S. Pless and W. C. Huffman, eds., *Handbook of Coding Theory*, Elsevier (1998).

[83] G. Polya, *Mathematics and Plausible Reasoning* (2 vols.), Princeton University Press (1954).

[84] G. Polya and G. Szegö, *Isoperimetric Inequalities in Mathematical Physics*, Ann. Math. Study No. 27, Princeton University Press (1951).

[85] G.-C. Rota, On the foundations of combinatorial theory I: Theory of Möbius functions, *Z. Warscheinlichkeitstheorie und Verw. Gebiete* **2** (1964), 340–368.

[86] A. Sali, Extremal theorems for submatrices of a matrix, Combinatorics (Eger, 1987), Colloq. Math. Soc. János Bolyai **52**, North-Holland, Amsterdam (1988), 439–446.
[87] K. Steiglitz and A. J. Bernstein, Optimal binary coding of ordered numbers, *J. SIAM* **13** (1965), 441–443.
[88] P. K. Ure, A study of (0,n,n+1)-sets and other solutions of the isoperimetric problem in finite projective planes, Ph.D. thesis, CalTech (1996).
[89] J. Vasta, The maximum rank ideal problem on the orthogonal product of simplices, Ph.D. thesis, University of California, Riverside (1998).
[90] B. L. van der Waerden, *A History of Algebra*, Springer-Verlag (1985).
[91] Virgil, *The Aeneid,* translated by C. D. Lewis, Doubleday (1953).
[92] D. L. Wang and P. Wang, Discrete isoperimetric problems, *SIAM J. Appl. Math* **32** (1977), 860–870.
[93] G. M. Ziegler, *Lectures on Polytopes*, Springer-Verlag (1995).

Index

The numbers in bold indicate where the term is defined.